Minimizing Uncertainty in the Design of Technical Systems

Yuri Spirochkin
International Consultancy and Analysis Agency 'Aviation Safety'
Aerospace Technology Department, Saint Petersburg, Russia
https://orcid.org/0000-0003-1362-9206

CRC Press
Taylor & Francis Group
Boca Raton London New York

CRC Press is an imprint of the
Taylor & Francis Group, an **informa** business
A SCIENCE PUBLISHERS BOOK

Cover illustration by Steve A Johnson (obtained through www.pexels.com)

First edition published 2027
by CRC Press
2385 NW Executive Center Drive, Suite 320, Boca Raton FL 33431

and by CRC Press
4 Park Square, Milton Park, Abingdon, Oxon, OX14 4RN

CRC Press is an imprint of Taylor & Francis Group, LLC

Library of Congress Cataloging-in-Publication Data (applied for)

ISBN: 978-1-041-08530-0 (hbk)
ISBN: 978-1-041-13595-1 (pbk)
ISBN: 978-1-003-67057-5 (ebk)

DOI: 10.1201/9781003670575

Typeset in Times New Roman
by Prime Publishing Services

Who knows what round the corner waits
To smite?
G.K. Chesterton, Vanity

Preface

Designers of airplanes, spacecraft and aerospace vehicles, nuclear power plants and other complex technical systems face the problem of uncertainty. They make decisions on the construction of a technical system that will ensure its required function (transportation of people or cargo, energy generation, etc.) with specified performance and safety characteristics. The feasibility and effectiveness of these solutions are substantiated on the basis of assumptions regarding the factors that will affect the system during operation and assessments of its response to these factors. Whereas the required function, performance and safety characteristics are clearly defined, the assumptions about the occurrence and intensity of operational factors, as well as the assessments of the system response, which are obtained through calculations or experiments, are subject to uncertainty – whether the designers realize it or not. This uncertainty ranges from stochastic variation in the values of known variables – system properties and factors acting under typical operating conditions – to the lack of knowledge about all possible adverse events during the life of the system and its behavior in such events.

Assumptions about the processes that are possible in the post-design stages of a system's life cycle and requiring assessment of its behavior, are based on the experience of operating prototypes (systems similar to the one being designed, including its experimental specimens), the results of research experiments, and theories that generalize experimental data. Any experience is limited by the conditions of its acquisition and does not provide reliable knowledge of what lies beyond these conditions. Unreliability of knowledge or its complete absence is one of the main causes of uncertainty.

Uncertainty is exacerbated by the impossibility to fully control negative human factors that may be inherent in the creators of the system and result in design errors, as well as those that may manifest themselves in those who will interact with the system during operation. Such manifestations include incorrect specification of future operating conditions, which is possible even with extensive experimental data, errors in calculations and experiments intended to substantiate design decisions, as well as unaccountable, marginal behavior by people, ranging from personnel violating established rules for operating and maintaining the system, to attackers who decide to use the system for their own purposes.

Modern vehicles, conventional or nuclear power plants, and processing facilities are saturated with energy and can contain flammable, toxic and radioactive technological substances. Transportation infrastructure and large buildings also possess significant potential energy. Any of such systems poses a potential hazard.

Uncertainty at the system design stage can lead designers to overlook potential extreme events, underestimate their intensity, and make unfounded assumptions about the system's ability to withstand them. If the intensity of an adverse event and the system's resistance, which deteriorates due to aging, combine in an unfavorable way, the system will collapse. The release of accumulated energy and the emission of hazardous technological substances can threaten human life and health, as well as the environment. An accident, and especially a catastrophe, can result in the complete loss of an expensive technical system. Clearly, the uncertainty that creates such a threat must be minimized.

The current philosophy of engineering design accounts for uncertainty using a set of theoretically sound and proven methods that, with proper quality control at the stages of manufacturing system elements and assembly (construction) of the system, as well as adherence to established operating rules, ensure reliability and safety over a long service life. However, accidents do occur from time to time, causing significant material damage, and even catastrophes accompanied by fatalities. These raise the following questions:

1. Did the system designers anticipate such events?
2. Did they do everything possible to prevent them?
3. Could the disaster has been prevented by using more sophisticated design methods?
4. Given that disaster-initiating events are characterized by uncertainty of nature, timing and intensity, is it possible to rule out such uncertainty in principle?

Answers to the first two questions can only be specific – relating to the technical system involved in the disaster; these are typically contained in investigation reports of such events. This book is an attempt to answer, at least partially, the third and fourth questions, which concern the general methodology of designing technical systems to prevent disasters in the future or to ensure their less severe consequences.

Numerous scientific publications are devoted to the study of uncertainty and the development of methods for managing it in design. These publications typically consider uncertain factors within the framework of probability theory, based on approximations of available statistics. When statistics are insufficient, more advanced theories are applied. I am an engineer who has worked for many years in high-tech industries – aerospace and nuclear. I have had the opportunity to participate in large scientific and technical programs such as the creation of the Soviet space transport system *Energia-Buran*, the design of the International Thermonuclear Experimental Reactor (ITER), the updating of the regulatory framework for the Russian nuclear energy, not to mention many smaller-scale research and development projects. I know that many of my colleagues do not trust complex mathematical constructions that cannot be verified in practice or using approximate "manual" estimates. The safety targets adopted in the aforementioned industries – an accident probability equal to 10^{-6} or a catastrophe probability equal to 10^{-7} – are unlikely to be confirmed experimentally, as there is simply insufficient statistical data. However, even when the statistical base is large and the probabilistic approximations seem reasonable, can one be sure that they cover all possible operational situations?

Traditional design approaches – both probabilistic and deterministic (in the latter, probabilistic characteristics are "hidden" in the factors of safety) – minimize uncertainty by relying on a priori assumptions about the operation stage. They enable designers to create reliable and safe systems for specified operating conditions, including predictable adverse events – so-called design basis accidents. Nevertheless, the very a priori nature of these approaches complicates their applicability to unforeseen situations where uncertainty exceeds the limits of predictability. Ensuring safety in such situations requires different design principles that would provide the system with "intelligence" to recognize emerging hazards, including previously unknown ones, in real time and the ability to respond flexibly to them. This book describes these principles, which reflect shifts in traditional philosophy of design toward minimizing uncertainty and thereby improving safety of the created system.

While working on this book, discussing its idea and some of provisions with colleagues, I was surprised to discover that not all of them, despite their extensive engineering experience, understood the problem of uncertainty. According to one colleague, uncertainty in design is commonplace, since creating a new machine always involves overcoming existing uncertainty, and such uncertainty requires no special study. Another colleague, who runs a private engineering firm, believed that if a customer's project contains uncertainty, it is best not to undertake it. The third opinion boiled down to doubts about the possibility of expressing safety and uncertainty in quantitative indicators.

When I began collecting material for this book, I myself did not fully understand the phenomenon of uncertainty – perhaps that is why the work on the manuscript, began in 2018, took about seven years and required considerable intellectual effort. Discussions with colleagues have strengthened my commitment to developing my own independent and informed judgment about uncertainty. Communication with Sergey A. Melnichenko, CEO of the International Consultancy and Analysis Agency 'Aviation Safety', drew my attention to new manifestations of uncertainty, such as the impact of space debris on aircraft flight safety and the introduction of artificial intelligence into aviation systems. Dimitar Popov, one of the founders of the Society of Aerospace and Nuclear Engineers, helped find materials on the theoretical foundations of neural networks and proved to be an invaluable interlocutor on these issues. Engineer Igor V. Uspensky performed the development and calculations that enabled the application of neural network technology in our joint research. Exchange of views with Nadezhda V. Yakimovich, an aviation psychologist, and familiarization with her publications, led to thoughts on how to take into account in design the uncertainty associated with critical situations in aircraft flight caused by specific mental states of pilots. This informational support and collaboration helped me in my work, and the results are reflected in the book. I am very grateful to my colleagues for their kind attention and efforts.

This book may differ from conventional scholarly publications, which aim to explore one or more related scientific problems and the methods developed to explore them. Scientific research involves in-depth study of a subject and leads to an extension of knowledge. Design, as opposed to research, is a synthetic activity, the ultimate goal of which is the creation of a material object, sometimes fundamentally new. To achieve this goal, designers consider a wide range of scientific and technical issues, organize research, and make practical decisions. This feature of design could not help but be reflected in the book – it touches on many scientific problems related to design, but generally without the in-depth exploration that is the domain of researchers.

The book is intended primarily for designers working in the aerospace industry and nuclear energy, including analysis engineers, test engineers, and project team leaders. It may also be of interest to specialists in other industries involved in the design and operation of potentially hazardous machines, facilities, and infrastructure. It may also be useful to researchers, engineering students, and postgraduate students – anyone who would like to expand their knowledge of the methodology of design and uncertainty inherent in this field.

Yuri Spirochkin, Saint Petersburg
February 27, 2026

Contents

1 Introduction

This introductory chapter defines the key terms used in the book, clarifies the entities under consideration and formulates the problem of uncertainty faced by designers of technical systems. It also describes the purpose of the book, its structure and the issues covered.

1.1 Key Definitions

1.1.1 Uncertainty

The Cambridge Dictionary gives the following definition of uncertainty[1]:

> *"a situation in which something is not known, or something that is not known or certain"*
> *and*
> *"the feeling of not being sure what will happen in the future".*

Another definition of uncertainty is proposed by the International Organization for Standardization: "state, even partial, of deficiency of information related to understanding or knowledge" (ISO 31073:2022).

These definitions contain such characteristic aspects of uncertainty as something *not known*, the corresponding *feeling* of a human trying to understand the situation or foresee the future, and deficient *understanding* or *knowledge*. Thus, a connection is established between uncertainty, on the one hand, and human consciousness and cognition, on the other. This helps to reveal the nature of uncertainty – apparently, uncertainty is inherent in human consciousness and, more generally, in the understanding of the world by living beings, in the accumulated knowledge, and in the behavior based on understanding and knowledge. Outside the consciousness, understanding, knowledge and behavior of living beings, in the objective world of inanimate matter, at least in macroscopic phenomena, there is no uncertainty. Therefore, uncertainty has a subjective origin.

The interactions of material bodies and physical fields occur in unambiguous accordance with the laws of physics – from Newtonian mechanics to Einstein's general theory of relativity. For example, if two inanimate bodies are moving towards each other and there is no obstacle between them, they will definitely collide. But if at least one of these bodies is a living being, then it can try to avoid the collision. That is, in the case of a living being, it seems that, along with the laws

[1] https://dictionary.cambridge.org/dictionary/english/uncertainty.

DOI: 10.1201/9781003670575-1

of physics, other factors also appear that lead to the "non-obligatory execution" of some of these laws – such factors are the understanding of the situation by this being, the innate instinct of self-preservation and the ability to act purposefully.

Einstein's famous statement that God "does not play dice" (Britannica, 2025) excludes uncertainty as a matter of chance. There remains in the world the uncertainty caused by ignorance (the lack of knowledge) about some objectively existing physical entity or interaction, and this uncertainty gradually recedes as the field of knowledge expands. Nevertheless, it seems impossible to completely eliminate it.

However, in the microworld described by quantum mechanics, uncertainty can exist at a fundamental level, according to Heisenberg's uncertainty principle, which states that there is a limit to obtaining precise values for certain pairs of physical properties of elementary particles and, therefore, to obtaining knowledge. In this case, it cannot be argued that this uncertainty is due solely to our insufficient understanding of nature and is subjective in nature, since it manifests itself in instrumental measurements, i.e., it has an objective aspect.

Thus, uncertainty is present in human life and activity at the macro level as a subjective phenomenon. At the same time, it manifests itself in quantum interactions in an objective way. The presence of uncertainty has a significant impact on such a specific area of activity as the design of technical systems, which encompasses both the macro and micro worlds. Consequently, uncertainty becomes an objective problem requiring research and solution.

1.1.2 Technical System

A technical system is a stationary or mobile artificial object that has the following distinctive features:

1) it is created to perform a specific function;
2) appropriate technologies are used in its creation, and
3) it includes a set of elements and connections between them.

Examples of technical systems are nuclear reactors, aerospace vehicles, industrial processing plants, infrastructure facilities, etc. The elements of each such system are parts of its structure, energy generators, engines or other units in which specific working processes occur, the control loop, safety devices, and so on. By using the term "element", we follow the terminology of *systems engineering* (ISO/IEC/IEEE 15288:2023; SEBoK, 2025) and Russian nuclear energy (NP-001-15), although this is somewhat inconsistent with the Western tradition of referring to machine parts as "components". We reserve the term "component" to designate parts of an element that, after its manufacture, form a single whole and are united by a common purpose, or for different but inseparable aspects of other entities and phenomena under consideration.

The elements themselves may be systems that perform individual functions to support the intended use of the system. For example, the function of structure is to bear the loads associated with the operation of the system. Thus, a modern complex technical system can be considered as a combination of nested systems, or subsystems, which in turn consist of smaller elements. The nesting of subsystems into larger systems can be multi-level, and the subsystems themselves can be unified for several applications. The number of such nesting levels and the unification of subsystems indicate the maturity of technologies for creating technical systems.

During operation, a technical system becomes part of an even more complex system – let us call it an *operational system* – the other parts of which are the human operator who controls the system and the environment in which the system works. A human operator is a specific link in the control loop of a technical system – his (or her) work is based on principles that are significantly different from those of machine components. The environment includes the natural conditions that physically affect the system, which is the first part; the people using this system, such as vehicle occupants or the public in the area where it operates, who may affect or be adversely affected by that operation, and other systems located near the system that can interact with it. Uncertainty can be present in each of the three above-mentioned parts: in the technical system itself, in the human operator, and in the environment.

The technical systems that surround us essentially form a "second nature". It is not surprising that in recent decades the quasi-biological term *life cycle* has come into use to characterize their existence.

This book focuses on technical systems related to two high-tech industries most familiar to the author – aerospace industry and nuclear energy. However, the basic provisions are also applicable to other areas of technology.

1.1.3 Design of Technical Systems

Design (or engineering design) is the first stage of the life cycle of any technical system, as shown in Figure 1.1. At this stage, its imaginary model (prefiguration) is created, which is not material and exists only in the documentation. The entire subsequent part of the life cycle can be called the *post-design* period. It consists of four stages: manufacturing of system elements; their supply to the construction site and construction (assembly) of the system; operation; and decommissioning.

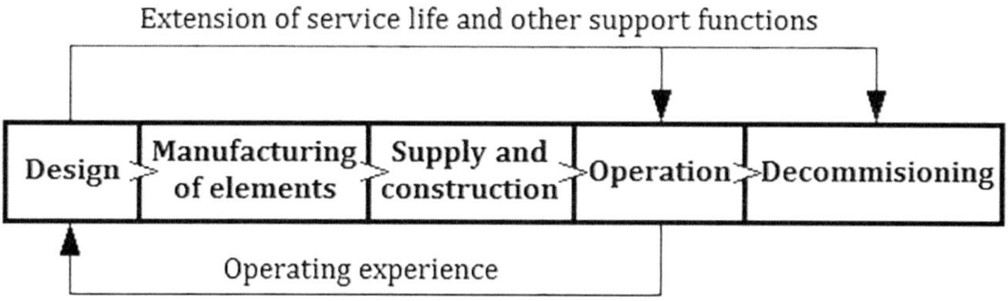

Fig. 1.1 Main stages of the life cycle of a technical system.

Dictionaries and standards provide a number of definitions of the term "design" in relation to human activity, but they do not fully and accurately reflect the content of the activity associated with the creation of technical systems. From the point of view of a designer – developer of technical systems – the following definition is preferable. Design of a technical system is a set of intermediate stages and types of work to determine its architecture, dimensions and properties as an ensemble of elements, and to prepare established technical documentation based on specifications, regulatory requirements and reference information. All this work is aimed at the material embodiment of the system and ensuring other post-design stages of its life cycle – mainly operation.

This definition reflects such important aspects as:

- the complexity of the design process, which includes intermediate stages and covers various types of work;
- its framework, or *design basis*;
- the result (technical documentation, the list and format of which are established by standards and (or) terms of the design contract);
- the purpose of the design (ensuring post-design stages of the life cycle of the system being developed).

The term "design basis" is used in nuclear energy, but has not gained widespread acceptance in other industries. It refers to the set of specifications, regulatory requirements, and reference information upon which a technical system is created. A more detailed definition can be found, for example, on the website of the U.S. Department of Energy (DOE, 2012).

The complexity of the design process is determined mainly by the fact that, from a methodological point of view, this process represents the solution of an inverse problem: determining the parameters of the technical system being created based on the specified (required) values of operational characteristics, i.e. determining the values of the arguments corresponding to the values of the function. Inverse problems, as a rule, do not have a single solution, i.e. they are characterized by uncertainty. They are usually solved by narrowing the search area for parameters to select one of the possible variants of a conceptual solution and then refining it through iterations.

The role of major iterations is played by intermediate stages, or phases, of design, the sequence of which corresponds to the increasing level of elaboration of design (constructive) decisions. Dividing the design process into such phases is aimed at managing complexity and early checking the feasibility of main decisions. In Western design culture, at least three phases are distinguished (Fielding, 1999; Pahl et al., 2007; Anderson, 2010):

- conceptual design;
- embodiment design (preliminary design);
- detail design.

Russian national standards establish four phases for the industrial design process: development of technical proposal, preliminary design, technical design, development of working design documentation (Spirochkin, 2023).

Within the second and third design phases, internal iterations are carried out, which include repeating two steps: 1) making a design decision – determining or updating previously adopted parameters of the system being created; and 2) assessing this decision through calculations or experiments, the results of which either substantiate its compliance with the design goal or reveal the need for changes. Initial design decisions can be based on prototypes and other reference information, generalized relationships between operational characteristics (the performance indicators) and parameters of the system, experience, formal or informal knowledge of designers. The assessments of the decisions are carried out by analysis engineers and experimental engineers specializing in various disciplines, the range of which depends on the class of the technical system under design. These assessments, which confirm the correctness of design decision, or lead to recommendations for changing it, implement the management of two main groups of indicators – one characterizes the system's performance, and the other – its safety.

The first group, performance targets, can be explicitly stated in the design contract or in specifications. They may also be implicit, inheriting the experience of using prototypes and implying the possibility of improvement during operation. Safety requirements (the second group) are mandatory – they are established in regulatory documents used in a given area of technology. Due to the mandatory nature of their implementation, it can be argued that safety actually takes priority over efficiency. The definition of safety of technical systems is given in Sect. 1.1.4. Ensuring safety is the main motive of this book.

Detailed design decisions substantiated by calculations and (or) experiments determine the final imaginary model of the system and are drawn up in the form of design documentation, which is used at the stages of manufacturing of elements, as well as supply and construction.

Design not only precedes other stages of the system life cycle, but is linked with them in other ways (see Figure 1.1). For example, the operation of existing systems can provide reference information on operating experience, which is valuable when creating new systems. Experience in using the first specimens of a serially produced system often acts as negative feedback for improving subsequent series. On the other hand, designers take part in activities carried out during operation: system modernization, extension of the designed service life, accident investigations, etc.

1.1.4 Safety, Risk and Related Terms

The Federal Law 'On technical regulation' in force in Russia defines safety as "a state in which there is no unacceptable risk associated with causing harm to the life or health of citizens, the property of individuals or legal entities, state or municipal property, the environment, the life or health of animals and plants" (No. 184-FZ, 2002). *Risk* is defined in this law as "the probability of causing harm ..., taking into account the severity of this harm".

The International Organization for Standardization gives a different definition of risk: "effect of uncertainty on objectives" (ISO 31073:2022). This definition implies a connection between risk, safety, as one of the objectives, and uncertainty, but does not reflect the fact that risk usually characterizes adverse events.

Thus, safety does not seem to have a quantitative measure of its own – it is expressed through risk, which can be quantified using, along with the probability of an adverse event, the severity of the harm it causes, or a combination of both. A common combined risk indicator, proposed in the early XVIII century by de Moivre, is the product of the severity of harm, or the magnitude of loss, and the probability of loss (Bernstein, 1996).

Adverse events that occur with technical systems can be divided into three categories according to the severity of the harm they cause: incident, accident, and catastrophe. An *incident* is a violation of normal operation of a system that is not serious enough to disable it. By *accident* we mean an adverse event that is usually unintentional and results in material damage and (or) environmental pollution and (or) human injury without fatality. The term *catastrophe* is used to describe an adverse event that results in loss of human life and (or) destruction and material losses on an extremely large scale.

Technical systems for which possible accidents are characterized by a very high risk, expressed in the severity of possible harm, and thus significantly affect safety, are classified as *safety-important* or *safety-critical* systems.

Safety of a technical system is mainly associated with the operation stage. However, ensuring safety is a task that must be addressed throughout the entire system life cycle, beginning with the design phase. This task can only be considered complete after the system is decommissioned, as damage can still occur during this final stage, when the system is dismantled and its elements are disposed of.

1.1.5 Multi-level Protection, or Defense in Depth

To ensuring safety of technical systems, the concept of *multi-level protection* against anticipated hazards is used. Originating in the military field, this concept is also called *defense in depth*. Outside this field, defense in depth has been most developed in nuclear energy. The main hazard associated with nuclear reactors, nuclear power plants (NPPs) and other nuclear facilities is the release of ionizing radiation beyond the structural boundaries of the facility. This hazard is countered by "a number of consecutive and independent levels of protection that would have to fail before harmful effects could be caused to people or to the environment" (IAEA, 2006). Some levels of protection are *physical barriers* formed by structural elements, while others are implemented through the use of *safety systems* and special organizational measures (IAEA, 2016). When one level of protection fails (one barrier is overcome), the next one comes into play. Properly organized multi-level protection ensures that no single failure due to technical malfunction or human error could lead to harmful effects and that a combination of failures threatening such consequences is of very low probability.

The number of independent levels of protection – primarily the number of physical barriers – can serve as a quantitative indicator of safety (another one, along with risk) achieved through defense in depth – the higher this number, the higher the safety. At modern NPPs of Russian design with the most common type of reactors – WWER[2] – the following are considered as physical barriers (NP-001-15):

1) fuel matrix that prevents fission products from escaping under the fuel element cladding;
2) fuel element cladding that prevents fission products from entering the primary coolant;
3) the boundary of the reactor coolant circuit, which prevents fission products from escaping into the volume under the hermetically sealed shell;
4) hermetically sealed shell (containment) that prevents the release of fission products into the environment, and
5) biological protection.

The same number of physical barriers is used in NPPs being built in other countries that are developing nuclear energy and maintaining an appropriate *safety culture* (IAEA, 2006).

The reliability of physical barriers, taking into account the expected degradation of their properties over the service life, is substantiated using calculations and tests at the creation of a nuclear facility, and is also supported through in-service monitoring and diagnostics, thereby minimizing uncertainty. These, as well as other technical and organizational measures, form functional levels of protection (NP-001-15; IAEA, 2016):

1) prevention of deviations from normal operation, as well as failures of elements important to safety, at all stages of the life cycle of a nuclear facility, starting with the design stage;
2) detection of deviations from normal operation conditions using special systems and devices and an attempt to take the deviations under control in order to prevent the occurrence of accident conditions and return the facility to a safe state;
3) if it is impossible to control deviations, then preventing the development of an accident due to the properties of *inherent safety* and (or) by engineered safety systems;

---

[2] Water-cooled Water Energetic Reactor corresponding to the international classification category "Pressurized Water Reactor" (PWR).

4) management of a severe accident[3] resulting from a failure of a previous level, using protective measures to mitigate its consequences and prevent further development threatening radioactive contamination;
5) mitigation of the harmful effects of radioactive releases if they occur during a severe accident, including measures to protect personnel, the population and the environment.

Planning of the relevant activities at the design stage of a nuclear facility includes defining specific requirements for instrumentation. It must provide the human operator with the information necessary to determine the actual state of the facility during an accident, make decisions on accident management and post-accident analysis. However, the need for rapid operator intervention must be minimized, otherwise "it shall be demonstrated that the operator has sufficient time to make a decision and sufficient time to act" (IAEA, 2016).

The large number of physical barriers to the main hazard, as well as a developed set of functional levels of protection that take into account the possibility of overcoming them, provide grounds for considering nuclear energy to be a very safe branch of technology.

Similar principles of protection are used in aerospace industry and other safety-critical areas of technology, but hazards to be considered may be specific to each class of technical systems – they may involve the uncontrolled release of energy possessed by the system and (or) explosive, flammable or toxic substances contained therein. Such release can occur when structural elements of the system lose their load-bearing capacity and collapse.

1.1.6 Structural Strength, Integrity and Reliability

The main pillar of ensuring safety of aerospace vehicles, nuclear reactors, as well as conventional transport, energy and industrial facilities that are potentially dangerous, including the approach through defense in depth, is maintaining the *strength* and *integrity* of their structures. The most comprehensive characterization of strength is likely contained in nuclear energy regulations. Strength is defined as the absence of the following limit states at any point in a structure when loads are applied (PNAE G-7-002-86):

1) short-term destruction (viscous and brittle);
2) creep failure under static loading;
3) plastic strain across the entire cross section of a structural part;
4) accumulation of maximum allowable creep strain;
5) cyclic accumulation of plastic strain that leads to not allowable changes in dimensions or quasi-static failure;

[3] Severe accident is characterized by the failure of physical barriers with an immediate threat of release of radioactive substances into the environment.

6) occurrence of macrocracks under cyclic loading;
7) loss of stability.

The limit states listed above are typical for both metals and other structural materials. However, for heterogeneous materials such as composites, additional limit states may be considered, e.g., delamination or separation of the filler from the matrix.

The term "integrity" is used along with strength; it characterizes the state of a structure as an ensemble of elements and takes into account the presence of detachable, including sealed, connections. For example, the loss of a small moving part of an actuator in a system control loop or the loosening of a bolted connection between a pressure vessel and a pipeline, resulting in a loss of tightness, may not affect the overall strength of the structure (up to a certain limit), but violate its integrity. Both can have a negative impact on safety – due to possible deterioration in system controllability or leakage of hazardous liquid or gaseous substances. Thus, integrity is a broader concept than strength. For different classes of technical systems, either strength or integrity of the structure under operating conditions is assessed, and different names for assessments of the same content may be traditionally used in different areas of technology.

Strength and integrity are expressed through strains, stresses, or displacements at points of a structure, i.e., through measurable variables. When any of these variables exceeds the allowable value, this means that the strength or integrity criterion corresponding to a certain limit state is not met. Therefore, they are quantitative indicators of the state of the system structure under load. The strength or integrity assessment completes the series of substantiations for constructive decisions carried out during the design of a technical system (Sect. 1.1.3). It either confirms safety from a structural mechanics perspective or identifies a safety deficit.

As noted above, in Sect. 1.1.4, safety does not have its own generally accepted quantitative measure, but is expressed through other measurable indicators. In addition to risk or the number of physical barriers, strength and integrity can be used in this capacity, as well as *reliability*. In Russian standardization documents, reliability is defined as "the property of an object to maintain over time the ability to perform the required functions in specified modes and conditions of use, maintenance, storage and transportation" (GOST 27.002-2015). Among the notes to this definition is the following statement: "reliability is a complex property which, depending on the purpose of the object and the conditions of its use, may include failure-free operation, maintainability, recoverability, durability, storability, availability, or certain combinations of these properties" (ibid). One of the most important components of this complex property from a safety perspective is *mechanical reliability*, which is a probabilistic quantity. It is discussed in Sect. 2.1.1.

1.2 The Problem of Uncertainty in the Design of Technical Systems

1.2.1 Range of Uncertainty and Impact on Safety

Safety of a technical system at the design stage is substantiated by assessing the strength (integrity) of its structure and mechanical behavior under static, dynamic, thermal loads, corrosion, erosion and other operational factors, which will hereinafter be referred to as "loads". Assumptions about these loads are based on operating experience with similar systems serving as prototypes and previously manufactured specimens – if the system is serially produced, as well as on the results of research experiments. Assessments are conducted through calculations using mathematical models and (or) testing of physical objects that, to varying degrees, reflect the properties of the system.

All components of the design process – the design decisions to be substantiated, assumptions about loads during operation, and assessments of the state or behavior of the system under these loads – carry some degree of uncertainty. It ranges from stochastic variation of actual properties of the system, arising during technological operations at the manufacturing of its elements and assembly (construction) or dispersion of values of known operational loads to the complete ignorance about all possible impacts at the operation stage and the behavior of the system in unforeseen situations. The left zone of this range, *aleatory uncertainty* (Figure 1.2), covers parameters and variables, as well as frequently occurring and predictable events of stochastic nature. The right zone, *epistemic uncertainty*, corresponds to adverse factors and events that are unknown at the time of design and therefore unpredictable. Between these extremes lies an intermediate zone, which includes impacts of known nature with unknown intensity, rare combinations of random events that may disrupt the operation of the system or make it unusable, errors in the analysis of the system behavior under ill-defined conditions, and variables that are difficult to estimate.

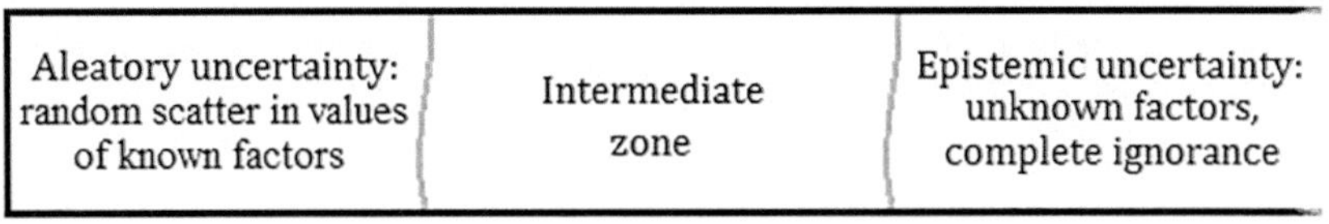

Fig. 1.2 Range of uncertainty in the design of technical systems.

Underestimation of loads in combination with incorrect data on actual properties of the system that determine its ability to withstand loads, errors in assessing the behavior of the system, and especially ignorance of any adverse factors, can lead to its collapse with the release of accumulated energy and hazardous substances contained within the system. In addition to the harm such collapse causes to people, property and the environment, the expensive system itself can be lost.

Thus, the uncertainty faced by the designers of a technical system creates a potential threat that becomes real under unfavorable circumstances during its operation. This threat is insufficient safety, or safety deficit. It reflects the impact of uncertainty on design, mentioned in Sect. 1.1.1, and, consequently, the negative impact on safety of the designed system. Because of all this, uncertainty is a serious problem for designers of technical systems, especially those that are of particular value, such as orbital and planetary space stations, deep space probes, aerospace vehicles, or those that are safety-critical, for example, nuclear facilities.

The design paradigm, developed over the last century and based on the use of statistical data from previous experience, the application of scientifically proven risk management methods, and the links between the design and post-design life cycle stages, including negative feedback (Sect. 1.1.3), ensures a level of safety generally acceptable to society. This paradigm has become generally accepted and traditional. However, even when operating highly reliable technical systems, accidents and catastrophes sometimes occur, revealing gaps in this paradigm. These emerge as manifestations of the hidden but inevitable uncertainty of the conditions and events that the system may encounter during operation.

1.2.2 Manifestations of Uncertainty: Examples of Accidents and Catastrophes

An example of manifestation of uncertainty, which combines the possibility of known hazardous events, random variations in the values of some factors, and the unknown intensity of others, is an emergency landing of an airplane. Such an accident should be analyzed at the design stage, but due to the a priori unknown values of key variables, its course and outcome are uncertain and depend largely on the actions of the crew. The causes of emergency landings may be, in particular, collisions with birds, when engines lose thrust, as happened on 15 January 2009, with *US Airways* Flight 1549. The Airbus 320 then ditched on the Hudson River near New York-LaGuardia Airport (ASN, 2009). Another similar case was the emergency landing of *Ural Airlines* Flight 178 on 15 August 2019, when after a dual bird strike and engine failure the crew landed the Airbus 321 in a cornfield with the landing gear retracted (ASN, 2019). In both of these cases, fatalities were avoided and the accidents did not turn into catastrophes thanks to the professional actions of the pilots.

Uncertainty in the form of unpredictable events outside design specifications and expected operating conditions has caused disasters such as:

- the crash of the *Concorde* supersonic liner, *Air France* Flight 4590, shortly after takeoff from Charles de Gaulle Airport, Paris, on 25 July 2000;
- the terroristic attack using two hijacked jets on the towers of the World Trade Center in New York on 11 September 2001;
- the large-scale nuclear accident at the *Fukushima Daiichi* NPP on 11 March 2011;

- the crashes of two Boeing-737-8 (MAX) airplanes – *Lion Air* Flight JT-610 near Jakarta on 29 October 2018 and *Ethiopian Airlines* Flight ET302 near Addis Ababa on 10 March 2019.

The designers of *Concorde* could not have foreseen that its fuel tank would be struck by a piece of tire that was torn when hitting a metal object lying on the runway. This impact damaged the tank, causing a fuel leak and a fire, after which the engines began to shut down and the plane crashed to the ground, killing 113 people on board and four on the ground (ASN, 2024b).

The Boeings hijacked by terrorists on 11 September 2001 were not protected from such actions, since no one foresaw them when designing these jets, and the towers of the World Trade Center were not designed to withstand an airplane impact. The jets were destroyed, fires broke out in the towers and they collapsed. The total number of victims exceeded 2,900 people (ASN, 2006a, 2006b).

The disaster at the *Fukushima Daiichi* NPP was initiated by a tsunami wave due to an earthquake in the ocean off the eastern coast of Japan. An unusually high (15-meter) wave overcame the protective wall (6 m high) and caused the flooding of the main power supply devices at all four nuclear reactors. The reserve diesel generators were also flooded. The interruption of power supply to the pumps of the cooling system led to the termination of coolant circulation and, as a result, to the destruction of cores in three of four reactors, explosions of the released hydrogen and partial collapse of the reactor buildings. A significant amount of radioactive material was released into the environment (IAEA, 2014). In nuclear community terminology, this event is classified as a nuclear accident. Fortunately, there were no fatalities, but because of enormous material damage, the event can be considered a catastrophe. This disaster demonstrated such difficult-to-predict aspects of uncertainty as the significant excess of the intensity of a known hazard (tsunami wave) over statistical data collected over the previous 100-year period, and a rare combination of random events – the failure of a number of physical barriers and safety systems.

Investigations into both Boeing 737-8 crashes (the first took the lives of 189 people, the second 157) revealed the same pattern: unpredictable, incorrect interaction between the recently installed Maneuvering Characteristics Augmentation System (MCAS), designed to assist pilots during stalls, and the pilot control actions. The main reasons for incorrect interaction were (ASN, 2022, 2024a):

- errors in the implementation of this system;
- inadequate certification procedure that failed to detect these errors: safety assessment of MCAS was carried out by the aircraft manufacturer itself, not by the regulator – the Federal Aviation Administration (FAA);
- lack of information on MCAS in the pilot manual.

It is worth noting that all these dramatic events (the list, unfortunately, could go on) occurred with serially produced technical systems and arose against the background of well-known operating conditions. Thus, even in a standardized life cycle, there is room for uncertainty. Uncertainty may manifest itself to an even greater degree in innovative projects that go beyond the existing experience, such as the creation of commercial suborbital transport, characterized by complex operating conditions, unconventional design and new structural materials, or long-term space missions using a combination of space and nuclear technologies. It inevitably follows that the problem of uncertainty is relevant and requires a solution. The obvious direction to seek a solution when complete elimination of uncertainty seems impossible is to minimize its level. The necessary conditions for minimizing uncertainty are a deeper understanding of its nature, quantitative assessment, and the possibility to control it, or, more broadly, manage it.

1.3 The Purpose of this Book, Its Structure and the Issues Covered

The purpose of this book is to

- explore the problem of uncertainty in the design of technical systems, focusing on safety-critical ones;
- analyze existing approaches to uncertainty management that reduce it, ensuring a level of safety acceptable for society, and reveal their advantages and shortcomings from the designer's perspective;
- propose new design approaches that promise to minimize the uncertainty associated with insufficiently known operating conditions and rare adverse events by leveraging modern technological advances, including artificial intelligence (AI), as well as implementing bionic (biologically inspired) ideas.

The book consists of nine chapters. This introductory chapter is followed by Chapter 2, which describes existing approaches to managing uncertainty in design. The traditional approach is presented, focusing on aleatory uncertainty and based on probabilistic approximations of statistical data obtained from previous experience. Attention is paid to situations where this approach faces difficulties due to the unreliability of the statistical base. Existing approaches are capable of reducing uncertainty to an acceptable level within a certain range of its manifestations, but they poorly cope with uncertainty associated with a lack of knowledge about the future operating conditions of the technical system being developed, i.e., epistemic uncertainty.

Chapter 3 attempts to delve deeper into the phenomenon of uncertainty in the design of technical systems to better understand it and overcome the shortcomings of existing approaches to its management. The sources of uncertainty, its structure, and components are identified. Some non-obvious issues related to uncertainty are also been revealed. As a result, it becomes possible to outline ways to minimize uncertainty over a broader range than previously possible, using currently available theoretical and experimental tools.

Chapter 4 examines ways to reduce uncertainty in design using theoretical tools – analytical methods. These include widening the object of analysis carried out when substantiating design decisions, refinement of mathematical models used in the analysis, reconstruction of the phenomenon being analyzed based on available fragments of information, and a number of others analytical tools.

The subject of Chapter 5 is approaches to reducing uncertainty based on obtaining new experimental data. Among these approaches are the refinement of probabilistic characteristics approximating the stochastic properties of a technical system and load parameters recorded in research experiments, the determination of the actual properties of system specimens and loading conditions during certification or other tests, applying the Bayesian method, pilot projects, as well as in-service monitoring and diagnostics. For each approach, the potential for minimizing uncertainty is revealed, and its inherent limitations are discussed.

Chapter 6 focuses on innovative approaches that are currently in the research and development stage. One of them is a methodology for analyzing design data to identify potential sources of epistemic uncertainty. Another approach is the concept of information support for the system life cycle, which constitutes the innovative core of life cycle management based on actual state. The next innovation considered is the active vibration diagnostics method, an advanced tool to determine the actual state of a structure by detecting hidden defects. A new design concept – Smart ProActive Resilient System (SPARS) – is also presented. It combines well-known approaches that form defense in depth against predictable adverse events with the capabilities of modern technologies and bionic ideas aimed at ensuring resilience to unforeseen dangers.

Chapter 7 provides a more detailed presentation of the SPARS concept using an example typical of aircraft design – the postulated case of emergency landing. An analysis of regulatory documents and existing design approaches reveals that there is significant uncertainty in this case and, as a result, a safety deficit that cannot be eliminated within the traditional design paradigm. To illustrate the capabilities of the SPARS concept for solving this problem, a hypothetical case – the emergency landing of the Soviet aerospace vehicle *Buran* – is considered. Along with previous developments, the current state of work in this area is described, characterized by the use of AI in the form of a neural network and the application of behavioral bionics principles.

Chapter 8 takes a closer look at AI to explore its potential for minimizing uncertainty in design. An overview of currently available AI elements from an engineering design perspective is given. Neural network technology, the most promising of these elements, is examined in more detail, and its applications to selected design problems are outlined. While the use of AI in design appears promising, when applied to safety-critical systems, it may introduce new risks. Opportunities for mitigating the risks associated with AI implementation are discussed.

Chapter 9, which concludes the book, attempts to capture the evolution of the engineering design landscape over time. It examines changes in the types of technical systems being created, the design methodologies being applied, and the regulatory approaches governing design activities. The impact of these changes on the uncertainty inherent in design is under analysis. It is hoped that the methods for dealing with uncertainty described in previous chapters will ensure its minimization, even in the face of emerging problems.

At the end of each chapter, there is a list of sources cited in the text – References. Among these sources, as well as the examples illustrating theoretical provisions, are some that date back several decades. This can be confusing for readers seeking fresh ideas and approaches. Of course, there is no new information in the past, but one can find something unknown to the current generation of specialists – either for some reason it was previously inaccessible, or, being known then, is now forgotten. The movement toward the future, associated with technological progress, is often accompanied by a loss of knowledge about past achievements. Without knowing at each specific interval of this trajectory what came before, we have no way to assess the effectiveness of our movement. Recognizing this problem and its potentially negative impact on the results of technical activity, the author, in addition to presenting new concepts, also sought to capture in the book the ideas and facts that characterize the achievements of previous generations of engineers and researchers, so that knowledge of their achievements will help those who come after them.

References

Anderson, J.D. Jr. (2010) *Aircraft performance and design*. Tata McGraw Hill, Delhi.

ASN (2006a) Aviation Safety Network – ASN Aviation Safety WikiBase. https://asn.flightsafety.org/wikibase/323228. Accessed July 04, 2025.

ASN (2006b) Aviation Safety Network – ASN Aviation Safety WikiBase. https://asn.flightsafety.org/wikibase/323229. Accessed July 04, 2025.

ASN (2009) Aviation Safety Network – ASN Aviation Safety WikiBase. https://asn.flightsafety.org/wikibase/321581. Accessed June 30, 2025.

ASN (2019) Aviation Safety Network – ASN Aviation Safety WikiBase. https://asn.flightsafety.org/asndb/319362. Accessed June 30, 2025.

ASN (2022) Aviation Safety Network – ASN Aviation Safety WikiBase. https://asn.flightsafety.org/wikibase/319474. Accessed July 04, 2025.

ASN (2024a) Aviation Safety Network – ASN Aviation Safety WikiBase. https://asn.flightsafety.org/wikibase/319547. Accessed July 04, 2025.

ASN (2024b) Aviation Safety Network – ASN Aviation Safety WikiBase. https://asn.flightsafety.org/wikibase/323465. Accessed July 04, 2025.

Bernstein, P.L. (1996) *Against the gods: the remarkable story of risk*. John Wiley & Sons, New York.

Britannica (2025) The Editors of Encyclopaedia Britannica. What did Albert Einstein mean when he wrote that God "does not play dice"? Encyclopaedia Britannica, February 13, 2025. https://www.britannica.com/question/What-did-Albert-Einstein-mean-when-he-wrote-that-God-does-not-play-dice. Accessed June 11, 2025.

Fielding, J.P. (1999) *Introduction to aircraft design*. Cambridge University Press, Cambridge, UK.

GOST 27.002-2015 (2016) Reliability in engineering. Terms and definitions. Interstate Standard. Standartinform, Moscow (in Russian).

DOE (2012) U.S. Department of Energy | Terms and Definitions. Design basis. https://www.directives.doc.gov/terms_definitions/design-basis. Accessed October 16, 2025.

IAEA (2006) IAEA Safety Standards Series No. SF-1. Fundamental safety principles. Safety fundamentals. International Atomic Energy Agency, Vienna.

IAEA (2014) IAEA report on human and organizational factors in nuclear safety in the light of the accident at the Fukushima Daiichi nuclear power plant. *International Experts Meeting*, Vienna, 21-24 May 2013. International Atomic Energy Agency, Vienna.

IAEA (2016) IAEA Safety Standards Series No. SSR-2/1 (Rev.1). Safety of nuclear power plants: Design. Specific safety requirements. International Atomic Energy Agency, Vienna.

ISO 31073:2022 (2022) Risk management – Vocabulary. International standard, Edition 1. International Organization for Standardization, Geneva.

ISO/IEC/IEEE 15288:2023 (2023) Systems and software engineering -- System life cycle processes. International standard. Edition 2. – International Organization for Standardization, Geneva / International Electrotechnical Commission / Institute of Electrical and Electronics Engineers.

No. 184-FZ (2002) Federalnyi Zakon 'O tekhnicheskom regulirovanii' ot 27.12.2002 No. 184-FZ (The Federal Law 'On technical regulation'), as amended on July 2, 2021. https://www.consultant.ru/document/cons_doc_LAW_4024. Accessed December 03, 2025 (in Russian).

NP-001-15 (2015) Federalnye normy i pravila v oblasti ispolzovaniya atomnoi energii. Obshchie polozheniya obespecheniya bezopasnosti atomnykh stantsii (Federal regulations and rules in the field of atomic energy use. General provisions for ensuring safety of nuclear power plants). Federal Environmental, Technological and Nuclear Supervision Service of Russia, Moscow (in Russian).

Pahl, G., Beitz, W., Feldhusen, J., Grote, K.-H. (2007) *Engineering design. A systematic approach*, 3rd edn. Springer, London.

PNAE G-7-002-86 (1989) Normy raschyota na prochnost oborudovaniya i truboprovodov atomnykh energeticheskikh ustanovok (*Rules for strength analysis of equipment and pipelines of nuclear energy facilities*). Gosatomenergonadzor of the USSR. Energoatomizdat, Moscow (in Russian).

SEBoK (2025) *Guide to the systems engineering body of knowledge*. Version 2.13. https://sebokwiki.org/wiki/Guide_to_the_Systems_Engineering_Body_of_Knowledge_(SEBoK). Released November 17, 2025. Accessed February 04, 2026.

Spirochkin, Y. (2023) *Human factors and design*. Springer, Singapore.

2 Existing Approaches to Uncertainty Management in Engineering Design and Other Areas

This chapter describes existing approaches to managing uncertainty in engineering design and some other technology-related areas. The traditional design approach is primarily applicable to aleatory uncertainty and is based on probabilistic approximations of statistical data obtained from previous experience. Analysis engineers involved in design typically use this approach implicitly, in the form of deterministic factors of safety that account for the dispersions of maximum load parameters and properties of the system characterizing resistance to loads. However, when experimental data is absent or insufficient to form a reliable statistical base, the probabilistic analysis faces difficulties. In such situations, the Bayesian method, based on subjective probabilities, can be applied. Along with this method, other approaches for managing uncertainty without using probabilities are described. Limitations of existing approaches that negatively impact the safety of designed systems are identified. These limitations necessitate a deeper understanding of the sources of uncertainty in design and the search for new ways to minimize it.

2.1 Traditional Design Approach Based on Probabilistic Approximation of Statistics

2.1.1 Separation of Uncertainty and Probabilistic Analysis of Its Aleatory Component

In order to manage uncertainty, more than 100 years ago in the socio-economic sciences it was proposed to divide it into two components, differing in genesis and measurability (Willett, 1901; Knight, 1921):

1) objective, or measurable uncertainty, expressed in the concept of risk, and
2) subjective, or immeasurable component, associated with the lack of knowledge about all possible outcomes of an event or all possible events in a given context.

Assigning an objective character to the first of these components does not correspond to our understanding of the subjective nature of the uncertainty as applied to macroscopic phenomena that seem random to human consciousness (Sect. 1.1.1), but the division itself coincides with the categorization of uncertainty as aleatory or epistemic (Sect. 1.2.1).

DOI: 10.1201/9781003670575-2

The first component, aleatory uncertainty, assumes that all possible variants of the realization of a random event are known and for each variant its probability P_i can be estimated ($i = 1, 2, ..., n$; where n is the number of variants). Probability estimates are obtained based on the approximations of available statistics on repeated events of the same type using suitable probability distribution laws. Such a statistically based probabilistic approach is widely used today in the natural sciences and engineering, and the methodological basis for taking into account aleatory uncertainty is *probability theory* (Ventsel, 1969).

Probability distribution laws are usually defined using probability density functions (PDFs). For a random continuous variable x (for example, maximum value of the load acting in a certain operating mode and subject to stochastic variation, the value of a material property or a geometric parameter of a structural element) the value of PDF, $p(x)$, is the probability that the value of x is in a small interval (a, b):

$$\int_a^b p(x)dx = P(a \le x \le b), \tag{2.1}$$

where ($a = x - \Delta$; $b = x + \Delta$ (Δ is a small increment of x).

Along with the PDF, the probabilistic approximation can take the form of a cumulative probability function (CPF):

$$P(x) = P(\chi \le x) = \int_{-\infty}^{x} p(\chi)d\chi, \tag{2.2}$$

where χ is a random variable that can take many values along the x axis.

Aleatory uncertainty is also inherent in some processes of changing loads over time during normal operation of a technical system (loading modes, or loading regimes), for example, the turbulent flow of a gas or liquid with which the elements of its structure interact. Such mechanical interaction can be approximated as a random process: $x(t)$. An important characteristic of a random process is the correlation function $K(t, t')$, which for each pair of arguments t and t' is equal to the correlation moment of its corresponding cross sections:

$$K(t,t') = M\left[\left(x(t) - \overline{x}(t)\right)\left(x(t') - \overline{x}(t')\right)\right], \tag{2.3}$$

where $\overline{x}(t)$ is the mean of the set of realizations of the random variable x at the moment t, and $\overline{x}(t')$ is the same at the moment t'.

If the process is stationary, i.e. $x(t)$ has a constant variance D, and for determining the correlation function, not the moments t and t' themselves are significant, but only the difference between them: $\tau = t' - t$, so that $K(t, t') = K(t, t + \tau) = K(\tau)$, then it can be described using the spectral power density (SPD):

$$G(\omega) = \frac{2}{\pi}\int_0^\infty K(\tau)\cos(\omega\tau)d\tau, \tag{2.4}$$

where ω is the circular frequency.

The variance for a stationary random process is:

$$D = \sigma^2 = \int_0^\infty G(\omega)d\omega, \tag{2.5}$$

where σ is the standard deviation of x.

Non-stationary random processes, the evolution of which over time does not allow analysts to assume the constancy of the variance and the independence of the correction function from the choice of a specific observation time t, can in some cases be represented as a sequence of time intervals, on each of which these conditions are approximately fulfilled. In other cases, more complex probabilistic methods are used for analysis (Bendat and Piersol, 1986). PDF, CPF, SPD and other probabilistic characteristics are obtained by processing statistical data accumulated over long-term experience in the relevant field of engineering or scientific discipline. In any of these approximations, the measure of aleatory uncertainty is probability.

The most common distribution laws used in engineering are the normal law, as well as the Weibull and Gumbel distributions. The normal, or Gaussian, distribution describes quite well the dispersion of mechanical properties and geometric parameters of structural elements manufactured in serial production conditions, as well as the stochastic variation of maximum load values acting in normal operating modes. The PDF for such a distribution is:

$$p(x) = \frac{1}{\sigma\sqrt{2\pi}} e^{-\frac{(x-\bar{x})^2}{2\sigma^2}}. \tag{2.6}$$

The Weibull distribution is applicable to approximating the dispersion of the fatigue strength characteristics of structural elements and carrying out reliability analysis. The Gumbel distribution is used mainly to describe extreme loads caused by adverse natural events: hurricanes, floods, etc. There are many other distribution laws that approximate various stochastic phenomena. For information on these areas of probability theory and their applications, the reader is referred to the relevant literature or the online handbook (NIST/SEMATECH, 2012).

Probabilistic mathematical models, constructed in analytical or numerical form, are used in disciplines such as decision analysis (Raiffa, 1968) and operations research (Morse and Kimball, 1951; Churchman et al., 1957; Ventsel, 1972). They are also tools of the methodology of probabilistic safety assessment (PSA) or probabilistic risk assessment (PRA) applied for safety-critical technical systems (NRC, 1990; INSAG, 1992; EPRI, 1995; IAEA, 2015; Boyer, 2018; Smith et al., 2024).

Given the necessary input data, probabilistic methods are well suited to analyze safety of the designed system by determining the mechanical reliability, R, of load-bearing elements of its structure. R is determined as the probability that the maximum parameter of the applied load corresponding to each possible limit state (peak value, loading duration, number of loading cycles, etc.), F, will not exceed the corresponding value of the load-bearing capacity, S:

$$R = P(F \leq S). \tag{2.7}$$

The load-bearing capacity of a structure is understood as its strength or integrity, expressed through load parameters. The safety analysis must cover all loading modes of the system during operation. The input data for such an analysis are the PDF of the maximum load parameter for each mode, $p(F)$, and the PDF of the corresponding value of the load-bearing capacity, $p(S)$, as shown in Figure 2.1.

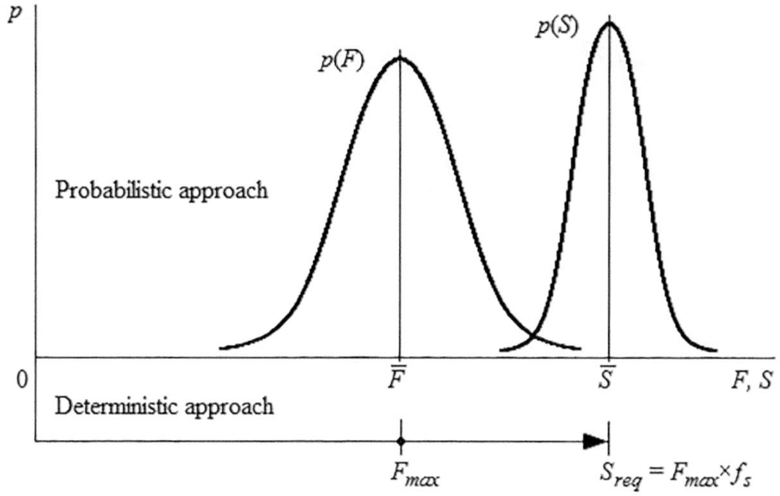

Fig. 2.1 Graphical representation of the probabilistic approach and its replacement by a deterministic one using the factor of safety.

2.1.2 Replacing Probabilistic Analysis with a Deterministic Approach

The probabilistic approach, although it underlies the traditional design paradigm, is applied directly only to a limited extent. One of the reasons for this may be that design is an activity of a predominantly synthetic nature: the result of design should be clearly determined geometric characteristics and physical and mechanical properties of all elements of the technical system being created. Another reason is the noticeable prejudice of designers against probabilistic methods. Most of the engineering community members known to the author view them with distrust.

One way or another, the probabilistic approach is implemented, as a rule, implicitly – through deterministic *factors of safety* (safety factors), *strength reserve factors* or *reliability factors*[1]. Using these coefficients, f_s ($f_s > 1$), the required load-bearing capacity of a structural element for each expected limit state, S_{req}, is increased relative to the corresponding maximum load parameter, F_{max} (see Figure 2.1):

$$S_{req} = F_{max} \times f_s. \tag{2.8}$$

The factors of safety – we will use this term – are established in regulatory documents as values that may vary for different system elements and operating modes. They take into account, in a hidden form, the dependence on known variances of maximum load parameters and the properties of structural elements characterizing their resistance to them, the type of approximating probability distribution law, and the required safety or mechanical reliability indicators. Assuming that the parameters of the load and load-bearing capacity are statistically independent normally distributed random variables, this dependence can be obtained as an explicit expression (Taylor, 1965; Makarevskii et al., 1975; Rzhanitsyn, 1978; Gladkii, 1982; Spirochkin, 2019):

$$f_s = \frac{1 + \sqrt{\gamma^2 (w_F^2 + w_S^2) - \gamma^4 w_F^2 w_S^2}}{1 - \gamma^2 w_S^2}. \tag{2.9}$$

In this equation, γ is the quantile corresponding to the required reliability value, w_F and w_S and are *coefficients of variation* of the maximum load parameter F and the load-bearing capacity S, respectively:

$$w_F = \frac{\sigma_F}{\overline{F}}; \; w_S = \frac{\sigma_S}{\overline{S}}, \tag{2.10}$$

where σ_F and $\overline{F}$ are the standard deviation and the mean of F; σ_s and $\overline{S}$ are the same characteristics of S.

Thus, for a probabilistic value of mechanical reliability, R, there is a corresponding deterministic value of the factor of safety, f_s, which performs the function of "distancing" the stress-strain state of a structural element from the possible limit state, ensuring guaranteed strength (integrity) under the intended operating conditions. The "distance" is measured in units of γ, i.e. is equal to the number of standards $\sqrt{(w_F^2 + w_S^2)}$, which fit into it. The higher the coefficients of variation w_F and w_S, the larger the factor of safety must be adopted to ensure the same required mechanical reliability and, accordingly, safety.

The coefficients of variation of the load-bearing capacity of structural elements are determined on the basis of experimental data. For a number of standardized configurations of elements and types of loading, that may lead to certain limit states,

[1] Different branches of technology use different terms, the essence of which is the same.

for example, for thin-walled panels under compression, which implies a possible loss of stability, these coefficients can be obtained directly in research experiments. In most cases, they are calculated using the coefficients of variation of mechanical properties (yield strength, ultimate strength, Young's modulus, Poisson's ratio, etc.) obtained in tests of material samples and experimentally determined coefficients of variation of dimensions of the supplied material assortment (sheets, bars, tubes, profiles and so on), taking into account the influence of processing technologies and joining of parts during the manufacture of elements. Probabilistic characteristics of mechanical properties, including coefficients of variation, are specific to each material and its supplied assortment; different assortment items are also characterized by corresponding coefficients of variation of dimensions (Okhapkin, 1970). The coefficients of variation of the mechanical properties of materials and geometric parameters observed in manufactured structural elements may depend on the technologies used. Methods of automatic processing and making joints by automatic methods are characterized by less influence on these coefficients than manual operations (Gladkii, 1982). In general, the values of the coefficients of variation of mechanical properties, geometric parameters and load-bearing capacity of structural elements reflect the technological level of production – the lower it is, the higher these values are.

An analysis of experimental data performed for aircraft structures shows that, depending on the stochastic variation of the load-bearing capacity, their elements can be divided into three classes (Gladkii, 1982):

1) the first (highest) class, which corresponds to the coefficients of variation $w_S < 0.06$;
2) the second (middle) class with the coefficients variation from 0.06 to 0.1 and
3) the third (lowest) class, for which the coefficients variation $w_S > 0.1$.

Experimental data are also used to determine the coefficients of variation of the maximum load parameters, w_F. Primary data, as a rule, are the results of measuring the kinematic parameters (velocities or accelerations) of prototype systems in loading modes characteristic of normal operating conditions, according to which the loads acting on structural elements can be found by calculations. The coefficients of variation of the kinematic parameters have been reliably established for a number of loading modes that can be characterized as stationary or close to stationary, e.g., rail or road transportation, hitting atmospheric turbulence, as well as for cases of short-term loading – such as aircraft landing. Assuming that the response of the structure in such loading conditions is directly proportional to the acting load, similar values of the coefficients of variation can be assigned for the maximum load parameters. For example, for loads arising during the transportation of structures and cargo on conventional railways and roads, the value of should be taken as no less than 0.35 (Gladkii, 1982). The coefficient of variation of the maximum load on the aircraft structure caused by clear-air atmospheric turbulence (outside clouds) is no less than 0.27, and during landing it can be equal, judging by measurements of vertical accelerations, to 0.533 (Taylor, 1965). In all the mentioned loading modes, the normal law of distribution of maximum loads can be assumed. For aircraft landings, the overwhelming majority of the studied cases (98%) correspond to it, and it is this main part of the statistical population that yielded the latest value of w_F. In 2% of landings that can be classified as "hard" (with high vertical velocities and accelerations), and in conditions of severe turbulence, characteristic of thunderstorms, the distributions of maximum loads may differ from the normal law, and the coefficients of variation have other values – most likely, smaller than those presented above due to an increase in the means (ibid). The probabilistic characteristics of such minor parts of the statistical populations, corresponding to extreme loads, determine the static strength of the structure.

The influence of the coefficients of variation w_F and w_S on the required value of the factor of safety, described by Eq. (2.9), is shown in Figure 2.2. The presented example of the dependence of f_S on these coefficients is obtained under the assumption that the required mechanical reliability of a structural element is equal to 0.9987, which corresponds to $\gamma = 3$.

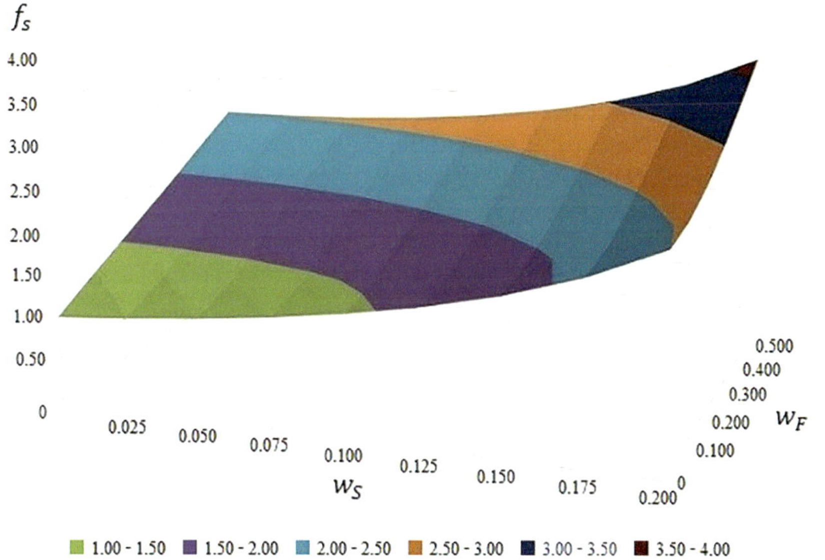

Fig. 2.2 The influence of the coefficients of variation w_F and w_S on the factor of safety f_S for the case of normal distribution of variables and the required mechanical reliability $R = 0.9987$.

In the airworthiness standards for transport category airplanes, the typical value of the factor of safety is 1.5 (CFR Part 25, § 25.303). The same value of f_S is used in the design of aerospace vehicles. According to Eq. (2.9) and the dependence shown in Figure 2.2, this value is applicable in cases where the coefficient of variation of the maximum load $w_F \leq 0.14$ and the coefficient of variation of the load-bearing capacity $w_S \leq 0.1$. The first condition means predictability and relatively low variability of loading modes, and the second corresponds, as noted above, to a high level of technology, implemented, as a rule, within the framework of stable serial production. For example, if we take $w_F = 0.1$, then with $w_S = 0.2$ (increased dispersion of properties or geometry parameters of structural elements, which is possible when introducing a new manufacturing technology), the factor of safety should have a significantly larger value: $f_S = 2.57$.

For other distributions of the load-bearing capacity and maximum load parameters, more suitable for some cases than the normal law, the factors of safety are determined by other, more complex expressions (Rzhanitsyn, 1978; Gladkii, 1982). However, the nature of the influence of the coefficients of variation on the values of f_S remains close to that considered above.

2.1.3 Measures to Manage Uncertainty

Since the factor of safety is fixed in regulatory documents, any change in the coefficients of variation of the load parameters or load-bearing capacity of the structural elements of the newly created system can lead to an actual change in the level of safety provided. A change in the coefficient of variation of the load parameters may result from changes in the operating conditions, whereas a change in the coefficient of variation of the bearing capacity is possible due to the use of new materials and (or) new manufacturing processes leading to specific variations in wall thickness of structural elements, eccentricity, shape deviations, etc. Therefore, to manage aleatory uncertainty in the design of safety-critical systems in order to maintain a high level of safety, along with the use of conservative factors of safety and statistically substantiated probabilistic distribution laws, measures are taken to ensure the stability of the dispersions of mechanical properties, geometry and loads. They include:

- the use of high-strength materials with small stochastic variation of strength characteristics, as well as the implementation of constructive decisions that limit the random scatter of geometric parameters of structural elements during their manufacture;
- the search for design options that can limit the impact of the dispersion of maximum load parameters on safety;
- verification of the applicability of existing probabilistic data when introducing new materials, manufacturing processes and structural configurations;
- in case of inapplicability, conducting experimental and theoretical research aimed at assessing the required probabilistic characteristics and the coefficients of variation[2];

[2]The purpose of this measure is not so much to ensure the stability of the dispersions, but to obtain correct data on them for specific design cases.

- more comprehensive consideration of uncertainty when substantiating design decisions through calculations;
- development of manufacturing, assembling and operating documentation that ensures limitation of the manifestation of uncertainty at the post-project stages of the life cycle.

The latter item in this list provides for quality management of all technological operations within the framework of standardized serial production – manufacturing of elements and assembly (construction) of the system, compliance with established regulations and rules for its operation with periodic monitoring, timely repair and replacement of system elements.

Such integrated management, covering the entire life cycle of the system, makes it possible to reduce aleatory uncertainty to a very low level – the target risk indicators in modern projects correspond to the probability of an accident $P_a = 10^{-6}$ and that of a catastrophe $P_c = 10^{-7}$. The question of the adequacy of these values remains open, among other circumstances due to the insufficient duration of operating experience of some systems, for example, NPPs (Spirochkin, 2023).

However, this framework for managing uncertainty proves inapplicable to its second, epistemic component, i.e. ignorance. If, when creating a technical system, there is reason to believe that poorly understood factors may be at play during operation, the traditional design approach assumes that knowledge about them can be acquired in several ways. The first involves conducting appropriate experimental or theoretical studies at the design phase. The second is the use of experimental specimens of the system preceding its serial production to collect missing data. The third is the installation of witness samples in a serial system, whose actual state can be observed during its operation. The applicability of these methods is determined by the nature of the unknown factors to be studied and the possibility of spending additional material and financial resources on the study within the framework of a specific project.

Thus, the traditional design approach:

- implies predictability of the operating conditions of the technical system being created and the characteristics of its resistance to loads based on available statistical data (from previous experience);
- includes the use of probabilistic approximations of this data (directly or implicitly – through the factors of safety);
- relies on the proper quality of all works performed during the manufacturing of system elements, its construction and operation to ensure minimal dispersions of actual system parameters and loads;
- thereby manages aleatory uncertainty and reduces it to an acceptable level;
- provides opportunities to work with epistemic uncertainty by expanding the boundaries of knowledge related to the system.

2.2 When the Traditional Approach Fails: "Black Swans", Poor Statistics, and "Unknown Unknowns"

2.2.1 Unexpected Events that Go Beyond Statistics, Fat-Tailed Distributions

The design approach described above enables the creation of technical systems with a level of safety generally accepted by society. Overall, it has proven itself over many years of creating, developing, and operating safety-critical systems, becoming not only a tradition but also a standard, shaping the current design paradigm. However, it has an inherent methodological flaw – even with statistics, predictions about the future based on past experience may be wrong because the potential of induction by which they are made is limited. Bertrand Russell criticized the reliance on inductive inferences as follows: "Domestic animals expect food when they see the person who feeds them. We know that all these rather crude expectations of uniformity are liable to be misleading. The man who has fed the chicken every day throughout its life at last wrings its neck instead, showing that more refined views as to the uniformity of nature would have been useful to the chicken. But in spite of the misleadingness of such expectations, they nevertheless exist. The mere fact that something has happened a certain number of times causes animals and men to expect that it will happen again. Thus our instincts certainly cause us to believe the sun will rise tomorrow, but we may be in no better a position than the chicken which unexpectedly has its neck wrung" (Russell, 1912).

This shortcoming of the inductive method is due to the epistemic uncertainty inherent in our knowledge of the world. Beyond the boundaries of knowledge lie unpredictable adverse events that are not covered by existing experience and statistics. The metaphor "black swan" has become generally accepted to characterize such events (Taleb, 2007). It is now applied to any unexpected and large-scale disasters including those mentioned in Sect. 1.2.2.

The inductive method also shows its weakness when applied to events whose nature is known but which may occur unexpectedly and with unpredictable intensity. Their uncertainty cannot be considered entirely epistemic, but their unexpectedness and destructive consequences produce almost the same psychological effect as completely unpredictable phenomena. Such an event can be considered the catastrophe occurred at the *Fukushima Daiichi* NPP in 2011. It was caused by the designers' underestimation of the maximum height of the tsunami wave based on available statistical data (IAEA, 2014b). The real problem, however, is that this variable, like magnitudes of many other natural phenomena, follow a power law distribution (Geist et al., 2009):

$$p(x) = Cx^{-\alpha}, \tag{2.11}$$

where x denotes in this case the maximum height of the tsunami wave; C and α are empirical parameters.

A variable distributed according to a power law has a definite mean only if $\alpha > 2$, and a finite variance if $\alpha > 3$ (Newman, 2005). The magnitudes of tsunamis, earthquakes, and the intensities of a number of other dangerous natural phenomena that can directly affect the safety of technical systems are characterized by exponent values $2 < \alpha < 3$ (ibid). Therefore, there is a possibility of events with extreme intensities that cannot be predicted. Any intensity limiting values can be exceeded, with corresponding fatal consequences.

The unexpectedness of such outliers beyond existing statistics and the unpredictable severity of their consequences raise doubts about the applicability of the probabilistic approach, and there appear proposals to change the traditional design paradigm based on it (Pariès, 2013). In this regard, it should be noted that for known and repeatable events such as tsunamis or earthquakes, the possibility of probabilistic assessment is, in principle, not in doubt. The criticism is rather directed at how to use these assessments, since the absence of the mean and variance breaks the approach to design based on the factors of safety (Sect. 2.1.2). In addition to the insufficient conservatism in the design estimates of the tsunami runup, another cause of the *Fukushima Daiichi* disaster was the incomplete implementation of the defense in depth concept (IAEA, 2014b; Spirochkin, 2023).

The uncertainty associated with known natural events whose magnitude is difficult to predict is aleatory rather than epistemic. The creator of the "black swan" metaphor, in his subsequent studies of unpredictable catastrophic events, focused on *fat-tailed probability distributions* that can approximate statistical series that include rare (or even single) extreme outliers (Taleb, 2020). There is no clear definition of a fat-tailed distribution. This category includes slowly decaying distributions (more slowly than, for example, the normal distribution), including those whose tails fall off according to a power law. A power law is the limiting case of a fat-tailed distribution. In fat-tailed distributions, extreme events corresponding to the tails are more probable than in a normal distribution. As a result, obtaining reliable estimates of the probability of such events is difficult. Risk estimates that assume a normal distribution for the random variable in question, usually turn out to be underestimated.

Despite the ability of fat-tailed distributions to describe mathematically rare extreme outliers that stand out sharply against the background of the main part of the statistics, the following questions arise:

1. If the magnitude of such a sudden single deviation differs significantly from the scale of other events in the series, does not this mean that some new unknown factor has come into play?
2. Can such a series be considered a statistical population (an immanent property of which is the qualitative homogeneity of events)?

It is the idea of a statistical population that allows us to obtain reasonable probabilistic estimates, not to mention the concept of a general population, on the basis of which the entire probabilistic methodology is developed. In these questions, the influence of epistemic uncertainty is undoubtedly evident.

2.2.2 Rare Events and the Bayesian Method

The extension of probabilistic approach to rare adverse events, for which statistical data are poor or absent, is possible on the basis of the Bayesian method (Ventsel, 1969). It operates with subjective probabilities that reflect not statistics, but the *degree of confidence* in one or other outcome of an analyzed situation, which are not predetermined. This method allows one to estimate the probability of one hypothesis being realized in comparison with another, alternative one, based on the latest relevant and significant evidence[3], using, for example, the following formula (Zlotnick, 1995):

$$P(H|A) = L(A, H) \times P(H), \tag{2.12}$$

where $P(H|A)$ is the probability of the outcome of the situation according to hypothesis H after considering the latest evidence A, i.e. the conditional probability of the realization of hypothesis in the presence of evidence A, the *a posteriori estimate*;

$P(H)$ is a preliminary estimate of the probability of realization of hypothesis H before considering the latest evidence (*a priori estimate* or value from previous analysis);

$L(A, H)$ is the *likelihood ratio*, the ratio of the conditional probability of the realization of evidence A when the hypothesis H is true, i.e. $P(A\,|H)$, to the total probability of the realization of this evidence, which also includes consideration of the alternative hypothesis.

This ratio can be determined as:

$$L(A, H) = \frac{P(A\,|\,H)}{P(A\,|\,H) \times P(H) + P(A\,|\,not\ H)[1 - P(H)]}, \tag{2.13}$$

where $P(A\,|not\ H)$ is the conditional probability that evidence A is not associated with hypothesis H, i.e. this evidence is realized for completely random reasons.

Thus, the Bayesian method allows designers to refine the subjective probability estimates they make in the face of insufficient statistics as new relevant data becomes available. An example of the application of such a method is the post-factum analysis of the terrorist attack on the World Trade Center in New York in 2001 (Silver, 2012). The prospects of it taking into account marginal manifestations of human factors in the design of technical systems (in the context of aleatory uncertainty) were considered by the author (Spirochkin, 2023). When applying the Bayesian method to the full spectrum of rare events that can affect the configuration and properties of the system under design, the above questions become even more

---

[3] Such evidence does not necessarily have to be "direct" – it only needs to have a non-random influence on the situation within the framework of a causal relationship.

important, since epistemic uncertainty only increases in the case of poor statistics and insufficient understanding of the nature of events.

2.2.3 *The Challenge of Epistemic Uncertainty*

In the causes of accidents and catastrophes, it can be difficult to separate the manifestation of aleatory uncertainty (quantified by probabilities) from epistemic uncertainty (immeasurable and non-quantifiable) – often, a combination of both is present. This complicates understanding of the impact of uncertainty on safety and, accordingly, uncertainty management. Perhaps the best comparative characterization of the two different categories of uncertainty was given by Donald Rumsfeld: "…There are known unknowns, that is to say we know there are some things we do not know. But there are also unknown unknowns – the ones we don't know we don't know"[4]. In his book, he described epistemic uncertainty as follows: "There are many things of which we are completely unaware – in fact, there are things of which we are so unaware, we don't even know we are unaware of them" (Rumsfeld, 2010).

The possibility of "unknown unknowns" appearing at post-design stages of the life cycle of a technical system, capable of leading to unpredictable consequences, represents a serious challenge for designers. The traditional design paradigm, well suited to managing the aleatory uncertainty associated with "known unknowns", is unable to cope with it.

2.3 Taking Uncertainty into Account without Relying on Statistical Data

To manage uncertainty when statistical data are unreliable or completely absent, attempts are made to structure it by identifying components that may not exactly coincide with aleatory uncertainty but nevertheless allow for quantitative assessment. One such attempt is to determine a component associated with known outcomes but unknown probabilities. The corresponding idea has its origins in socio-economic studies (Keynes, 1937; Stirling, 2007; Aven and Steen, 2010), and then has found its place in engineering (Beer et al., 2013; Patelli et al., 2014; Saunders et al., 2015; Beer et al., 2016). This component can be classified as aleatory uncertainty if estimates of *subjective imprecise probabilities* are feasible. Having such probabilities compatible with the common probabilistic approach, one can further refine them according to the Bayesian method or process them based on random set theory using probability boxes (p-boxes), Dempster-Shafer structures (Dempster, 1967; Shafer, 1976), etc. This approach narrows the area of epistemic uncertainty that cannot be managed.

[4] Donald Rumsfeld said this in 2002, when he was US Secretary of Defense, at a press briefing at the Pentagon; quote from (Rumsfeld, 2010).

When it is possible to make assumptions about some outcomes of an event, but their entire set is unknown (as are the probabilities), and there are only ideas about the boundaries within which the outcomes lie, then researchers try to identify other components in uncertainty. Among these components the following are considered: *ambiguity* (Lempert et al., 2003; Stirling, 2007; Saunders et al., 2015), *fragmentary or dubious information, imprecise variables* (Beer et al., 2016). Each of these, although falling primarily into the category of epistemic uncertainty, differs from complete ignorance by the presence of certain logical constructions – limited and controversial, individual opinions – including alternative ones, and assessments of varying degrees of validity. Several approaches are used to handle such incomplete, vague and unreliable information. Thus, in relation to ambiguity, a common psychologically based practice is ambiguity aversion (also known as uncertainty aversion) – this is reflected in the fact that people prefer known risks to unknown ones (Ellsberg, 1961) and reject decisions associated with the latter. In conditions of conflicting alternative opinions, controversial and contradictory assessments (sometimes reflecting only peculiar aspects of human factors), handling with the corresponding component of uncertainty is possible through appropriate project management methods.

In areas where safety is of particular importance, the *precautionary principle* is used (Stirling, 2007; EUR-Lex, 2017). It encourages caution and detailed research before embarking on innovations that could have disastrous results. Along with this principle, the method of postulating the *worst-case scenario* is used (Schneier, 2010; Capitanescu and Wehenkel, 2013). A similar method is *extreme case analysis*, which involves identifying events that lead to the worst-case behavior of a system (Patelli et al., 2014).

Among the practical methods of taking uncertainty into account, it is also worth mentioning the *duplication* of system elements, which is widely used in engineering. Duplication through redundant, or reserved, elements (including those working on different principles) has become standard in safety-critical systems. Although the effectiveness of this method can be substantiated using probability relationships – see, e.g., Pavlov and Razgulyaev (2017), it is intuitive and does not require any statistical estimates or theoretical assumptions.

There are also analytical tools that can process ill-defined information for which probability estimates are impossible or impractical:

- fuzzy logic (Zadeh, 1965; Yen, 1999; Singh et al., 2013; Gupta, 2021);
- interval analysis (Moore, 1966; Rao and Berke, 1997; Moore et al., 2009; Modares and Mullen, 2014; Sotiropoulos and Tserpes, 2022; Gong et al., 2023);
- sensitivity analysis (Haug et al., 1986; Patelli et al., 2014; Madni and Bahill, 2021; Blumsack, 2023), and
- heuristic strategies (Nielsen and Molich, 1990; Nielsen, 1994; Yilmaz et al., 2011; Pierre, 2015).

In each of the above attempts to structure uncertainty, as in its basic dichotomy, there is an epistemic component in the form of ignorance that cannot be managed in the given circumstances. It is closely intertwined with the controllable aleatory component, and this relationship complicates the task of minimizing the overall uncertainty in a certain project.

2.4 Organizational Ways of Dealing with Uncertainty

2.4.1 Introducing the Main Ways

When considering approaches to managing uncertainty in the design of technical systems, it would be unfair not to mention ways of dealing with it that are implemented not through scientific and technical methods, but rather at the organizational level. The potential for managing uncertainty at this level is determined primarily by two empirically observable and apparently fundamental circumstances (Knight, 1921): 1) uncertainty is less in groups of similar cases than in single instances, and 2) the perception of uncertainty and the decision-making aimed at reducing it depend on differences among human individuals who face it.

The first circumstance is used in uncertainty management by *grouping* or *consolidating* multiple similar cases. This way of mitigating the threat associated with the uncertainty of expected negative events, of which only a small part is realized in reality and causes harm, is embodied in such an institutional mechanism as *insurance*. Institutionalization means that insurance is carried out by relevant organizations whose activities are provided for by law and regulated by the state. Obviously, insurance is feasible primarily under conditions of aleatory uncertainty.

The second circumstance is exploited through *specialization*: the selection of individuals capable of assuming responsibility for decisions made under conditions of uncertainty, or the separation of the risky line of business associated with innovation from the stable and predictable part of the organization's activities. This approach is not limited to the domain of statistically predictable conditions, but also extends to epistemic uncertainty.

A brief description of these two approaches is presented below, in Sects. 2.4.2 and 2.4.3. Beside these, in activities with a high level of uncertainty of the outcome, a way of *uncertainty and risk decoupling* is applicable. The possibility of such decoupling follows from the distinction noted above (see Sect. 2.1.1) between the categories of risk and subjective uncertainty. For the former, "the distribution of the outcome in a group of instances is known (either through calculation *a priory* or from statistics of past experience)", while for the latter, "it is impossible to form a group of instances, because the situation dealt with is in a high degree unique" (Knight, 1921). However, even in a unique situation, we can usually estimate the magnitude of potential losses. This way is typical, for example, for business focused on creating fundamentally new commercial products, or for innovations in the field of defense and deterrence (Cocron and Aronhime, 2022). The practical implementation of the described decoupling is based on the intuitive (but generally confirmed) assumption of an inverse relationship between the probability of adverse events and the magnitude of losses. Actions whose success is questionable are undertaken in an area characterized by low losses but a high expected probability of occurrence. By performing these actions on a "small scale", it is possible to explore uncertainty. This approach is related to the *pilot project* concept, which is discussed in Sect. 5.4.

2.4.2 Insurance

Insurance is widely used in the transport sector, including civil aviation (Wells and Chadbourne, 2007). Commercial aircraft or private jet insurance includes a combination of risk protection measures: public liability, passenger liability and aircraft hull insurance. An aircraft's susceptibility to damage depends on many factors related to its design, overall configuration, age, and piloting. All these factors must be taken into account during underwriting – the process of selecting and assessing (pricing) risks faced by prospective policyholders and presented to the insurer. This process places high demands on the qualifications of underwriters. They must also be aware of new hazards that arise as flight speeds increase, technology improves, and piloting becomes more sophisticated.

In the second half of the XX century, insurance became common practice in nuclear energy – see, for example, NRC (2025). In the United States, insurance coverage for accidents at NPPs is regulated by the Price-Anderson Act of September 2, 1957. It covers bodily injury, illness, disease or subsequent death, property damage, and losses, including reasonable living expenses for evacuees. Claims may include any adverse event, including those caused by theft or sabotage, during the transportation of nuclear fuel to a reactor site; during the storage of nuclear fuel or waste at a site; during reactor operation, including the discharge of radioactive substances; and during the transport of irradiated nuclear fuel and nuclear waste from the reactor.

In Russia, according to the Federal Law 'On the use of atomic energy' (No. 170-FZ, 1995), liability insurance for nuclear damage is mandatory for all nuclear industry enterprises. Similar legal acts have been adopted in other countries developing nuclear energy. In addition to national legislation, the international Convention on Supplementary Compensation (CSC), developed under the auspices of the International Atomic Energy Agency, is in effect. The CSC provides an additional amount of liability coverage for nuclear incidents (IAEA, 2014a).

The history of space insurance dates back to 1965, when the first spacecraft, the Intelsat 1 geostationary satellite ('*Early Bird*'), was insured. Currently, all operators conducting space launches from US territory are required by law to insure their liability for each mission (Zisk, 2022). However, US regulators do not require liability insurance from satellite operators. In contrast, liability insurance is mandatory for satellite operators in the EU and the UK.

In the initial stages of space insurance development, approaches similar to those used in insuring aviation assets were used. However, space activities are characterized by significant technological novelty, unique production processes, launch infrastructure, and flight operations, as well as specific risks. The profile of these risks is changing with the expansion of spaceflight, including long-range interplanetary missions, suborbital transportation, space tourism, and so on. Relevant new approaches were required, and space insurance has become a distinct area from aviation insurance, with its own challenges, regulatory framework and methodology (Malinowska, 2017a).

Insurance premiums for space assets are high, reflecting the rapid pace of innovation and the significant risk of space technologies, as well as the limited amount of statistical data. They can be 10 to 20 times higher than insurance premiums common in aviation (Zisk, 2022). The insurance premium depends on the successful launch history, confirming the reliability of the launch vehicles of a particular type. It may be lower in relative terms than for a new launch vehicle type with shorter or no history. The same applies to new types of satellites. Insurers generally consider the failure of an experimental satellite to be more likely, leading to a corresponding increase in the premium.

Insurance premium levels also depend on the overall situation in the global space industry. Space has recently been viewed as a key risk area. This is due to the increasing frequency of space missions, a significant portion of which are commercial launches to low-Earth orbit, the accelerated growth of orbital debris, and the underdevelopment and outdated nature of international regulation of safety in space activities. Consequently, underwriting costs and the time required for underwriters to analyze and assess risks are increasing.

This is one reason why satellite system operators often forego insurance. The share of insured missions has never exceeded 60%, but by 2025, it has fallen to 20% (Forrester, 2025).

In general, for innovative industrial activities involving the creation and use of complex technical systems of a new type, characterized by extreme operating conditions and high potential losses, the risk profile (as a subject of insurance) may be either poorly defined or completely unknown. It may change as the scope of innovation expands and advanced technical systems are designed that have no analogues. This situation is typical for both the aerospace industry and nuclear engineering, including the currently developing projects for nuclear-powered spacecraft and extraterrestrial manned stations powered by nuclear reactors. Insuring innovative projects is associated with significant difficulties, including the need for highly qualified underwriters (who must possess the necessary technical knowledge to assess risks). Clearly, in this case, insurers must take an individual approach to each project (Malinowska, 2017b). Specialization and the separation (decoupling) of uncertainty and risk, two other approaches briefly introduced in Sect. 2.4.1, can also be in demand.

In concluding our characterization of insurance as one of the main approaches to dealing with uncertainty at the organizational level, we note the following:

- this approach has an institutional form and is supported by the state;
- its application can be rationally justified in the presence of statistical data, i.e., in the case of aleatory uncertainty, whereas under conditions of epistemic

uncertainty it lacks such justification and inevitably carries a significant subjective component;
- this approach does not reduce the uncertainty faced by the insured, but is aimed at compensating for their losses in the event of an unfavorable outcome by shifting monetary liability to the insurer;
- thus, the insurance premium paid by the insured is the price of objectified uncertainty, or, in other words, a monetary measure of uncertainty accepted by members of society.

2.4.3 Specialization

Specialization, as an organizational approach to dealing with uncertainty, can be implemented, as noted in Sect. 2.4.1, either in the personal dimension or on a business scale. In the former case, it involves the selection of individuals, for example, leaders of project teams, design organizations, scientific institutes, and enterprises, by higher-level managers or through independent career advancement. Such individuals must possess the ability to anticipate the outcome of uncertain situations, make decisions aimed at a favorable outcome, and ensure the achievement of set goals. Certain personal qualities are necessary for occupying leadership positions and subsequent success in activities characterized by uncertainty. These qualities include the following (Knight, 1921):
- enhanced foresight ability;
- the ability to "read" human nature and anticipate the behavior of others – often in contrast to scientifically based judgments;
- the ability to judge the means to achieve goals, plan actions, and assess the possibilities of adapting to predictable situations;
- the ability to implement plans;
- the trust that individuals feel in their judgments and their own abilities (regardless of the truth of these judgments and assessments of their abilities);
- a *conative attitude* toward the situation being judged (often associated with a preference for original hypotheses and for accepting rather than avoiding uncertainty, coupled with a predisposition to "believe in luck").

Let us illustrate specialization in dealing with uncertainty in design activities with a historical example. It concerns the Soviet lunar project (1960-1970), which envisaged the creation of the N-1 rocket and a lunar module for landing humans on Earth's satellite. The Chief Designer and head of this project was S.P. Korolyov[5]. While compiling a list of the main technical parameters of the lunar module, determining its future design, the question of the landing gear arose, and the mechanical properties of the lunar surface were completely unknown at the time.

[5] Korolyov, Sergei Pavlovich (1907-1966), was one of the main creators of Soviet rocket and space technology; Chairman of the Council of Chief Designers; Chief Designer of the R-7 rocket, which first lifted a man into space on April 12, 1961; head of the design organization now known as JSC S.P. Korolyov Rocket and Space Corporation *Energia*.

The opinions of planetologists diverged: some believed that the surface of the Moon could be similar to granite rocks, others that it is loose dust several tens of meters thick, and so on. At one technical meeting, the Chief Designer put an end to the protracted debate in his characteristically decisive manner: "Most scientists are inclined to believe that the soil on the Moon is hard. Like granite, or limestone, or pumice – these are just details – but hard. And the arguments of those who support this point of view seem more convincing than the arguments of the opponents". He paused and concluded decisively: "That's what we'll assume". The Chief Designer wrote in his notebook: "The Moon is hard," signed it "S. Korolyov," dated it, tore the page out of the notebook, and handed it to the engineer who would be directly supervising the design of the lunar module (Gallai, 1985). Korolyov's confidence in the hardness of the lunar soil was soon confirmed by the lunar landings of American and Soviet unmanned spacecraft. However, at that dramatic moment of design decision-making, one cannot help but credit the Chief Designer's intuition and his determination to assume responsibility in the face of uncertainty.

Specialization on a business scale is embodied in a tendency that accompanies industrial development in directions characterized by high uncertainty or a significant speculative component: the elements of industry that are associated with such directions are separated, as noted in Sect. 2.4.1, from stable and predictable elements and become self-sustained organizations. A typical example of this form of specialization is the existence in the USSR aviation industry of special design organizations (OKB[6]) that created new types of aircraft, while serial production of aircraft of previously developed designs was assigned to "regular" aircraft factories operating "according to plan".

The features of specialization, the second of the main approaches to dealing with uncertainty at the organizational level, are as follows:

- this approach is intended primarily for use in conditions of epistemic uncertainty;
- it is based on the manifestations of human factors in various dimensions, covering both the personal qualities of an individual and the activities of a team or organization;
- it does not reduce the level of epistemic uncertainty, but merely guides the activities of a human collective along a path chosen by the responsible person based on subjective judgments;
- due to a significant degree of subjectivity, this path can lead to both success and failure.

2.5 Generalized Assessment

The above overview of existing approaches to uncertainty management in the design of technical systems demonstrates the combination of two main components in this

---

[6]OKB is an abbreviation for the Russian term "Opytno-Konstruktorskoe Byuro" (experimental design bureau).

phenomenon: aleatory and epistemic, and a traditional design approach based on a quantitative assessment of the first of them. Quantification of aleatory uncertainty is possible due to the availability of statistics covering the operating experience of previous systems similar to the one being designed, and approximation of these statistics by appropriate probabilistic models. This component of uncertainty is taken into account directly by using such models in design analyses aimed at safety assessment or, most often, indirectly – through the factors of safety established in regulatory documents of a particular industry. Thus, the traditional design approach implements risk management, which is also carried out at the post-design stages of the system life cycle. Risk reduction methods, supported by a statistical base and quality control of all work, ensure safety of technical systems at a theoretical level corresponding to reliability, as the probability of the system functioning without accidents or catastrophes, no less than

$$R = 1 - max\{P_a, P_c\} = 1 - 10^{-6} = 0.9\,(6). \tag{2.14}$$

This level seems acceptable to society, although the possibilities for checking it are limited by a number of circumstances, including insufficient experience in operating some systems.

These methods are not applicable to managing epistemic uncertainty, or ignorance. The traditional design approach involves reducing ignorance through theoretical research or experiments conducted during the design stage, as well as collecting data during the operation of early, experimental specimens of the system prior to its serial production, or by using witness samples. Such measures undoubtedly contribute to increasing safety of technical systems, although this increase cannot always be quantified. Moreover, they cannot completely eliminate epistemic uncertainty.

The reliance on statistics, while a solid foundation for the traditional design approach, also determines its vulnerability. First, the inductive method, which promises reliable estimates about the future based on past experience, may lead to errors due to factors unknown at the time of design and coming into play in the post-design stages of the life cycle of a technical system. Second, the statistical data required to obtain probability estimates may not exist. Finally, even if statistical data exists, but the magnitudes of adverse events and, correspondingly, the maximum load parameters are described by fat-tailed probability distributions, then no load value, no matter how large, considered in the design can guarantee safety, since it may be exceeded as a result of an unpredictable extreme event. In such cases, the entire system of the factors of safety collapses.

The application of the traditional design approach to rare adverse events that are not sufficiently reflected in statistics or are absent from it is possible using the Bayesian method. This allows designers to work with data that represents subjective probability estimates (i.e. judgments), keeping in mind that they can be refined as new relevant data are obtained. The accepted estimates, despite their subjective nature, can be used in calculations performed through probability theory methods. However, if the new evidence needed to produce the Bayesian posterior estimate is only realized as a result of a disaster, the cost of refinement becomes too high. In many cases, the Bayesian method can be practically useful because it extends the scope of design analysis, but by translating judgments into a probabilistic dimension, it also inherits the inextricable connection of such a probabilistic methodology with aleatory uncertainty, only one of the components of uncertainty.

Uncertainty accounting methods that do not rely on statistics complement the traditional design approach and make it possible to compensate for its shortcomings associated with the lack of reliable statistics or to circumvent the limitations of the Bayesian method; however, they do not form any holistic alternative concept, but are applied rather *ad hoc*, i.e., depending on the specific situation. And these methods do not provide the management of epistemic uncertainty. Such of them, as sensitivity analysis or heuristic strategies, only allow designers in some cases to detect its manifestations in advance.

The constant accumulation of experience in the operation of technical systems, its theoretical generalization and transformation into knowledge undoubtedly contribute to the reduction of epistemic uncertainty. However, the area of technologies is constantly expanding, uncontrolled changes are taking place in the world, and the volume of epistemic uncertainty does not seem to decrease. In design, it is closely intertwined with aleatory uncertainty. Their connection complicates the minimization of both components.

Long-term operation of a technical system in an environment with changing climatic, meteorological, hydrological, and possibly seismic characteristics can be interpreted rather as a non-stationary random process than as a stationary one. In such a process, the mean and variance of the maximum load parameter corresponding to each limit state change over time. Due to the aging of materials and wear of structural parts under the influence of adverse operational factors, the average values of the strength characteristics decrease over time and their coefficients of variation may increase. Changes in the probability characteristics of the maximum load parameters and the bearing capacity of the structural elements affect the actual mechanical reliability of the technical system, reducing it.

The introduction of new materials, new structural configurations, as well as new technologies for manufacturing elements, assembling systems and maintaining their operability during operation leads to a change in the probabilistic characteristics of the resistance of new generation systems to loads. With the development of AI, which is being used in unscrewed vehicles and unmanned energy or production facilities, the very principles of ensuring reliability are changing. Manifestations of human factors in the life cycle of a technical system are also transformed, and in an ambiguous way. At the same time, safety requirements are becoming more stringent. All these processes determine not only the need to adjust the design approach to ensuring safety and, at a minimum, the factors of safety. A deeper understanding of the sources of uncertainty in design is needed, and finding new ways to minimize it is highly relevant.

References

Aven, T. and Steen, R. (2010) The concept of ignorance in a risk assessment and risk management context. *Reliability Engineering & System Safety*, 95(11):1117–1122.

Bendat, J.S. and Piersol, A.G. (1986) *Random data. Analysis and measurement procedures*. John Wiley & Sons, New York.

Beer, M., Ferson, S. and Kreinovich, V. (2013) Imprecise probabilities in engineering analyses. Departmental Technical Reports (CS). 734. https://scholarworks.utep.edu/cs_techrep/734. Accessed October 17, 2024.

Beer, M., Ferson, S. and Kreinovich, V. (2016) Do we have compatible concepts of epistemic uncertainty? Departmental Technical Reports (CS). 1006. https://scholarworks.utep.edu/cs_techrep/1006. Accessed July 15, 2024.

Blumsack, S. (2023) Decision analysis without probabilities. EME 801: Energy Markets, Policy, and Regulation. Lesson 6 – Cost-Benefit Analysis and Decision Making. The Pennsylvania State University. https://www.e-education.psu.edu/eme801/node/577. Accessed May 27, 2025.

Boyer, R.L. (2018) Probabilistic risk assessment (PRA): analytical process for recognizing design and operational risks. NASA Johnson Space Center. Safety & Mission Assurance. https://ntrs.nasa.gov/api/citations/20200001598/downloads/20200001598.pdf. Accessed June 25, 2025.

Capitanescu, F. and Wehenkel, L. (2013) Computation of worst operation scenarios under uncertainty for static security management. *IEEE Transactions on Power Systems*, 28(2):1697–1705.

CFR Part 25. Code of Federal Regulations/Title 14/Chapter I/Subchapter C/Part 25 – Airworthiness Standards: Transport Category Airplanes. https://www.ecfr.gov/current/title-14/chapter-I/subchapter-C/part-25. Accessed July 11, 2025.

Churchman, C.W., Ackoff, R.L. and Arnoff, E.L. (1957) *Introduction to operations research*. John Wiley & Sons, New York; Chapman & Hall, Ltd, London.

Cocron, A. and Aronhime, L. (2022) Risk, uncertainty and innovation. *NATO Review*. Opinion, analysis and debate on security issues, 14 April 2022. https://www.nato.int/docu/review/articles/2022/04/14/risk-uncertainty-and-innovation/index.html. Accessed May 30, 2025.

Dempster, A.P. (1967) Upper and lower probabilities induced by a multivalued mapping. *Annals of Mathematical Statistics*, 38(2):325-339.

Ellsberg, D. (1961) Risk, ambiguity, and the Savage axioms. *Quarterly Journal of Economics*, 75(4):643-669.

EPRI (1995) PSA applications guide. TR-105396. Final report. Electric Power Research Institute, Palo Alto, CA.

EUR-Lex (2017) Precautionary principle. European Union. EUR-Lex | Access to European Union law. https://eur-lex.europa.eu/legal-content/EN/TXT/?uri=LEGISSUM:precautionary_principle. Accessed July 16, 2025.

Forrester, C. (2025) Report: Satellite insurance sector suffering. *Advanced Television*, October 28, 2025. https://www.advanced-television.com/2025/10/29/report-satellite-insurance-sector-suffering. Accessed October 29, 2025.

Gallai, M.L. (1985) S chelovekom na bortu *(With a man on board)*. Sovetskii pisatel, Moscow (in Russian)

Geist, E.L., Parsons, T., Brink, U.S.T., and Lee, H.J. (2009) Tsunami probability. In E.N. Bernard and A.R. Robinson (eds.) *The Sea*, Volume 15: Tsunamis, Chapter 4, pp. 93-135. Harvard University Press, Cambridge, MA.

Gladkii, V.F. (1982) Veroyatnostnye metody proektirovaniya konstruktsii letatelnogo apparata *(Probabilistic methods for designing spacecraft structures)*. Nauka, Moscow (in Russian)

Gong, J., Wang, X., Lv, T. (2023) A credible interval analysis method for uncertain structures under nonprobabilistic framework. *Computer Methods in Applied Mechanics and Engineering*, Volume 404, Article 115833.

Gupta, A.K. (2021) Fuzzy logic and their application in different areas of engineering science and research: a survey. *International Journal of Scientific Research in Science and Technology*, 8(2):71–75.

Haug, E.J., Choi, K.K. and Komkov, V. (1986) *Design sensitivity analysis of structural systems*. Academic Press, Inc., Orlando, FL.

IAEA (2014a) Convention on Supplementary Compensation for Nuclear Damage. IAEA – International Atomic Energy Agency. https://www.iaea.org/topics/nuclear-liability-conventions/convention-supplementary-compensation-nuclear-damage. Accessed December 07, 2025.

IAEA (2014b) IAEA report on human and organizational factors in nuclear safety in the light of the accident at the Fukushima Daiichi nuclear power plant. *International Experts Meeting*, Vienna, 21-24 May 2013. Annex B. International Atomic Energy Agency, Vienna.

IAEA (2015) Probabilistic safety assessment. Basic Professional Training Course. Module VII. International Atomic Energy Agency, Vienna.

INSAG (1992) Safety Series No. 75-INSAG-6. Probabilistic safety assessment: INSAG-6: a report by the International Nuclear Safety Advisory Group. International Atomic Energy Agency, Vienna.

Keynes, J.M. (1937) The general theory of employment. *The Quarterly Journal of Economics*, 51(2):209–223.

Knight, F.H. (1921) *Risk, uncertainty and profit*. Houghton Mifflin Company, The Riverside Press Cambridge, Boston and New York.

Lempert, R.J., Popper, S.W. and Bankes, S.C. (2003) Shaping the next one hundred years: New methods for quantitative long term strategy analysis. MR-1626-RPC. The RAND Pardee Centre, Santa Monica, CA.

Madni, A.M. and Bahill, A.T. (2021) Handling uncertainty in engineered systems. In G.S. Metcalf, K. Kijima, H. Deguchi (eds.) *Handbook of Systems Sciences*, pp. 1167–1199. Springer Nature, Singapore.

Makarevskii, A.I., Korchyomkin, N.N., Frantsuz, T.A. and Chizhov, V.M. (1975) Prochnost samolyota. Metody normirovaniya raschetnykh uslovii prochnosti samolyota *(Aircraft strength: methods for standardizing the design criteria of aircraft strength)*. Ed. by A.I. Makarevskii. Mashinostroenie, Moscow (in Russian).

Malinowska, K. (2017a) *Space insurance: international legal aspects* (Aerospace law and policy series). Kluwer Law International BV, The Netherlands.

Malinowska, K. (2017b) Risk assessment in insuring space endeavours: a legal approach. *Air & Space Law*, 42(3):329–348.

Modares, M. and Mullen, R.L. (2014) Dynamic analysis of structures with interval uncertainty. *Journal of Engineering Mechanics*, 140(4):04013011.

Moore, R.E. (1966) *Interval analysis*. Prentice-Hall, Hoboken, NJ (Prentice-Hall series in automatic computation).

Moore, R.E., Kearfott, R.B. and Cloud, M.J. (2009) *Introduction to interval analysis*. Society for Industrial and Applied Mathematics, Philadelphia, PA.

Morse, P.M. and Kimball, G.E. (1951) *Methods of operations research*. The Technology Press of Massachusetts Institute of Technology and John Wiley & Sons, New York.

Nielsen, J. and Molich, R. (1990) Heuristic evaluation of user interfaces. *Proc. ACM CHI'90 Conference "Human Factors in Computing Systems"*, Seattle, WA, 1-5 April, 1990, pp. 249–256.

Nielsen, J. (1994) Enhancing the explanatory power of usability heuristics. *Proc. ACM CHI'94 Conference "Human Factors in Computing Systems"*, Boston, MA, April 24-28, 1994, pp. 152–158.

NIST/SEMATECH (2012) NIST/SEMATECH e-Handbook of statistical methods. https://www.itl.nist.gov/div898/handbook. Accessed July 10, 2025.

No. 170-FZ (1995) Federalnyi Zakon 'Ob ispolzovanii atomnoi energii' ot 21.11.1995 No. 170-FZ (The Federal Law 'On the use of atomic energy', edition dated December 26, 2024. https://www.consultant.ru/document/cons_doc_LAW_8450. Accessed December 03, 2025 (in Russian)

NRC (1990) Severe accident risks: an assessment for five U.S. nuclear power plants. NUREG-1150. Final summary report. Division of Systems Research, Office of Nuclear Regulatory Research, U.S. Nuclear Regulatory Commission, Washington, DC.

NRC (2025) Backgrounder on nuclear insurance and disaster relief. United States Nuclear Regulatory Commission. NRC Library, Last reviewed/updated: August 06, 2025. https://www.nrc.gov/reading-rm/doc-collections/fact-sheets/nuclear-insurance.html. Accessed December 07, 2025.

Newman, M.E.J. (2005) Power laws, Pareto distributions and Zipf's law. *Contemporary Physics* 46(5):323-351.

Okhapkin, S.O. (ed.) (1970) Veroyatnostnye kharakteristiki prochnosti aviatsionnykh materialov i razmerov sortamenta *(Probabilistic characteristics of the strength of aviation materials and of dimensions of assortment)*. Mashinostroenie, Moscow (in Russian).

Pariès, J. (2013) Why a paradigm shift is needed. *IAEA International Experts' Meeting on Human and Organizational Factors in Nuclear Safety in the Light of the Accident at the Fukushima Daiichi Nuclear Power Plant*. Vienna, Austria, May 21-24, 2013. Conference ID: 45441 (I2-CN-212). https://www-pub.iaea.org/iaeameetings/IEM5/IEM5_Jean%20Paries_Dedale_France.pdf. Accessed July 13, 2025.

Patelli, E., Alvarez, D.A., Broggi, M. and de Angelis, M. (2014) Uncertainty management in multidisciplinary design of critical safety systems. AIAA. https://core.ac.uk/download/pdf/80778445.pdf. Accessed October 22, 2024.

Pavlov, I.V. and Razgulyaev, S.V. (2017) Confidence estimation of reliability indices of the systems with elements duplication and recovery. *Mathematics & Mathematical Modelling*, No. 06, pp. 32–39 (in Russian).

Pierre, R.S.Q. (2015) Heuristics in design: a literature review. *Procedia Manufacturing*, Volume 3, pp. 6571–6578.

Raiffa, H. (1968) *Decision analysis. Introductory lectures on choices under uncertainty*. Addison-Wesley Publishing Company, Reading, MA.

Rao, S.S. and Berke. L. (1997) Analysis of uncertain structural systems using interval analysis. *AIAA Journal*, 35(4):727–735.

Rumsfeld, D. (2010) Known and unknown: authors note. From pages XIII-XVI. The Rumsfeld Papers. December 2010. https://papers.rumsfeld.com/about/page/authors-note. Accessed July 13, 2025.

Russell, B. (1912) *The problems of philosophy*. Home University Library, 1912; Oxford University Press paperback, Oxford (UK), 1959.

Rzhanitsyn, A.R. (1978) Teoriya rascheta stroitelnykh konstruktsii na nadezhnost *(Theory of calculating building structures for reliability)*. Stroiizdat, Moscow (in Russian).

Saunders, F.C., Gale, A.W. and Sherry, A.H. (2015) Conceptualising uncertainty in safety-critical projects: a practitioner perspective. *International Journal of Project Management*, 33(2):467–478.

Schneier, B. (2010) Worst-case thinking. Schneier on Security, May 13, 2010. https://www.schneier.com/blog/archives/2010/05/worst-case_thin.html. Accessed October 22, 2024.

Shafer, G. (1976) *A mathematical theory of evidence*. Princeton University Press, Princeton and London.

Silver, N. (2012) *The signal and the noise. Why so many predictions fail – but some don't*. Random House, New York.

Singh, H., Gupta, M.M., Meitzler, T., Hou, Z.-G, Garg, K.K., Solo, A.M.G and Zadeh, L.A. (2013) Real-life applications of fuzzy logic. *Advances in Fuzzy Systems*. Volume 2013, Article ID 581879.

Smith, C.L., Blanc, K.L. and Mandelli, D. (eds.) (2024) *Risk-informed methods and applications in nuclear and energy engineering: modelling, experimentation, and validation*. Academic Press.

Sotiropoulos, D.G. and Tserpes, K. (2022) Interval-based computation of the uncertainty in the mechanical properties and the failure analysis of unidirectional composite materials. *Mathematical and Computational Applications*, 27(3):38.

Spirochkin, Y.K. (2019) Bezopasnost rossiiskikh AES s tochki zreniya inzhenera-mekhanika *(Safety of Russian NPPs from the viewpoint of mechanical engineer)*. SUPER Izdatelstvo, Saint Petersburg (in Russian).

Spirochkin, Y. (2023) *Human factors and design*. Springer, Singapore.

Stirling, A. (2007) Risk, precaution and science: towards a more constructive policy debate. Talking point on the precautionary principle, *EMBO Reports*, 8(4):309–315.

Taleb, N.N. (2007) *The Black Swan. The impact of the highly improbable*. Random House, New York.

Taleb, N.N. (2020) *Statistical consequences of fat tails: real world preasymptotics, epistemology, and applications* (Technical Incerto Collection). STEM Academic Press.

Taylor, J. (1965) *Manual on aircraft loads*. Pergamon Press, Oxford; London.

Ventsel, E.S. (1969) Teoriya veroyatnostei *(Probability theory)*, 4th edn. Nauka, Moscow (in Russian).

Ventsel. E.S. (1972) Issledovanie operatsii *(Operations research)*. Sovetskoe radio, Moscow (in Russian).

Wells, T.A. and Chadbourne, B.D. (eds.) (2007) *Introduction to aviation insurance and risk management*. Third edition. Krieger Publishing Company, Malabar, FL.

Willett, A.H. (1901) *The economic theory of risk and insurance*. The Columbia University Press (Reprint for the S.S. Huebner Foundation for Insurance Education, University of Pennsylvania, by Richard D. Irwin, Inc., Homewood, Illinois, 1951).

Yen, J. (1999) Fuzzy logic – a modern perspective. *IEEE Transactions on Knowledge and Data Engineering*, 11(1):153–165.

Yilmaz, S., Daly, S.R., Seifert, C.M., Gonzalez, R. (2011) A comparison of cognitive heuristics use between engineers and industrial designers. In J.S. Gero (ed.) *Design Computing and Cognition '10*, pp. 3-22. Springer, Dordrecht, Netherlands.

Zadeh, L.A. (1965) Fuzzy sets. *Information and Control*, 8:338–353.

Zisk, R. (2022) The space insurance landscape. *Payload*, October 31, 2022. https://payloadspace.com/the-space-insurance-landscape. Accessed October 28, 2025.

Zlotnick, J. (1995) Bayes' theorem for intelligence analysis. In H.B. Westerfield (ed.) *Inside CIA's private world: declassified articles from the Agency's internal journal*, 1955–1992, Chapter 20, pp. 255–263. Yale University Press, New Haven, CT.

3 A Deeper Understanding of Uncertainty from the Designer's Perspective

It is becoming increasingly clear that existing approaches to managing uncertainty in the design of technical systems cannot effectively address its manifestations beyond accumulated experience and knowledge. The management of such manifestations, which combine aleatory and epistemic components, is complicated by insufficient understanding of the structure of this combination in each specific case, the sources of uncertainty and their impact on the design results. This chapter attempts to clarify these issues through a deeper dive into the phenomenon of uncertainty. It is expected that a more detailed understanding of uncertainty will facilitate the search for further ways to minimize it in design.

3.1 Sources of Uncertainty in the Design of Technical Systems

Existing approaches to managing uncertainty in the design of technical systems, reviewed in Chap. 2, do not guarantee safety in conditions of insufficient or absent statistical data, required to substantiate design decisions. This means that these decisions remain potentially unreliable if adverse events beyond the scope of existing experience are possible during operation. To analyze such events and assess the impact of unknown factors on safety, a deeper understanding of the sources of uncertainty is necessary, enabling designers to find ways to minimize the negative impact.

Uncertainty in design arises from the following causes (they can also be called sources of uncertainty):

- stochastic variation of the properties and geometry of structural elements, as well as maximum loads across an ensemble of possible implementations, which leads to the fact that the data characterizing the load-bearing capacity and behavior of a certain structure at the operation stage cannot be known precisely, but contain inevitable errors;
- hidden imperfections of the system – its structure, subsystems, control loop, etc. (poor design decisions, anomalies or defects[1]), and undetected

[1] A *defect* differs from an *anomaly* only by its regularized characteristics: it must belong to a known, specific type of imperfection (for example, a macrocrack in a structural element), and its quantitative parameters must not be lower than the established limits.

DOI: 10.1201/9781003670575-3

discrepancies in the operating documentation (gaps or inadequate provisions); they can manifest themselves unexpectedly and lead to violation of normal operation;

- a fundamental feature of any system: the emergence of new properties after connecting its elements – some of the emerging properties may be undesirable and non-obvious until certain, previously unknown conditions occur;
- possible uncontrolled impacts and changes in operating conditions of the system, capable of going beyond the limits provided for by the design, up to the occurrence of an emergency situation, when the further state of the system and its behavior become difficult to predict;
- human factors embodied in operating personnel and other people involved in the life cycle of the system or present in the area of its operation, taking into account that human behavior is largely characterized by unpredictability;
- inaccuracies in the experimental and theoretical (computational) substantiations of the design decisions due to incomplete similarity of prototypes, research make-ups and small-scale physical models to the properties of the created system, imperfections of mathematical models, defects in software and errors in the simulation results;
- the possibility of unknown factors acting during the operation of the system, which entails a priori unpredictability of its state and behavior.

The uncertainty associated with the first source is a consequence of the influence on the properties and geometry of the structural elements, as well as on the maximum load parameters, of many difficult-to-control deviations in the processes of manufacturing the elements, assembling the system and its operation. Such influence can additionally include changes caused by the aging of materials and wear of structural elements or other trends that appear during long-term operation. The uncertainty generated by the first source can be almost entirely classified as aleatory. Aleatory uncertainty analysis is well developed within the traditional design approach, which is based on probabilistic approximations of available statistical data on the properties of structural elements and load parameters, as described in Chap. 2. This analysis is performed either by probabilistic methods (when probabilistic safety assessments are performed) or by using deterministic factors of safety (in most cases when substantiating design decisions).

With sufficient accuracy for practical purposes, it can be assumed that the stochastic variation in the mechanical properties, geometric parameters and, accordingly, the load-bearing capacity of structural elements of traditional configuration, manufactured using standardized technologies, corresponds to the law of normal distribution (Gladkii, 1982). For typical operating modes of existing technical systems, probabilistic characteristics were also determined that approximate the loading parameters of their structural elements in the best or, at least, acceptable way. However, when introducing new materials, unusual structural configurations or advanced manufacturing and assembly processes, as well as when implementing operating modes that are different from those used previously, problems with the applicability of known probability characteristics may arise. Within the traditional design paradigm, these problems are solved by organizing, in parallel with the direct design work, experimental and theoretical studies aimed at updating the existing statistical base taking into account innovations.

In general, the uncertainty caused by the first source and the corresponding risk of failure of system elements are reduced by:

- high quality of all design and production work during the creation of a technical system, which ensures that the stochastic variation of key parameters remains within known limits;
- compliance with established operating rules, implementation of operational control, maintenance and repair in the operation stage, which prevents a reduction in mechanical reliability;
- timely decommissioning of the system when the degradation of its properties due to aging can lead, according to available statistical data, to insufficient reliability.

The level of development and the power of probabilistic methods determine the attractiveness of their application to other sources of uncertainty. Therefore, attempts are being made to apply probabilistic assessments to take into account hidden imperfections in the system (the second item on our list), which remain undetected during factory quality control, acceptance and certification tests (McEvily, 2002; Sukhoruchenkov and Okorokov, 2014). The likelihood that not all poor design decisions, anomalies and defects will be detected before commissioning is due to limited control or testing, insufficient technical characteristics of the tools used and the negative influence of human factors embodied in the personnel performing the relevant work. This influence covers a wide range, from the inability of designers to anticipate the full variety of potential imperfections and prevent their occurrence in the system, to the lack of skill and attention of inspectors and testers in identifying hidden defects.

The history of technology is replete with examples of design decisions that seemed promising when a system was developed, but turned out to be imperfect in practice. One such solution was implemented in the first version of the Russian solar sailing spacecraft, designed to participate in the race to Mars in 1992, which was planned to commemorate the 500th anniversary of Columbus discovery of America. The race did not take place due to financial problems faced by its organizers, but the Russian solar sail was created, and another potential use was found for it – as a sunlight reflector to illuminate dark areas of the Earth's surface (Spirochkin and Saprykin, 2021). This idea was tested on February 4, 1993, during the space experiment *Znamya*-2. The solar sail was a 20-meter-diameter disk formed by eight sectors of thin aluminum-coated polymer film, as shown in Figure 3.1. It was launched folded into low-Earth orbit on the *Progress* M-10 transport spacecraft and then deployed.

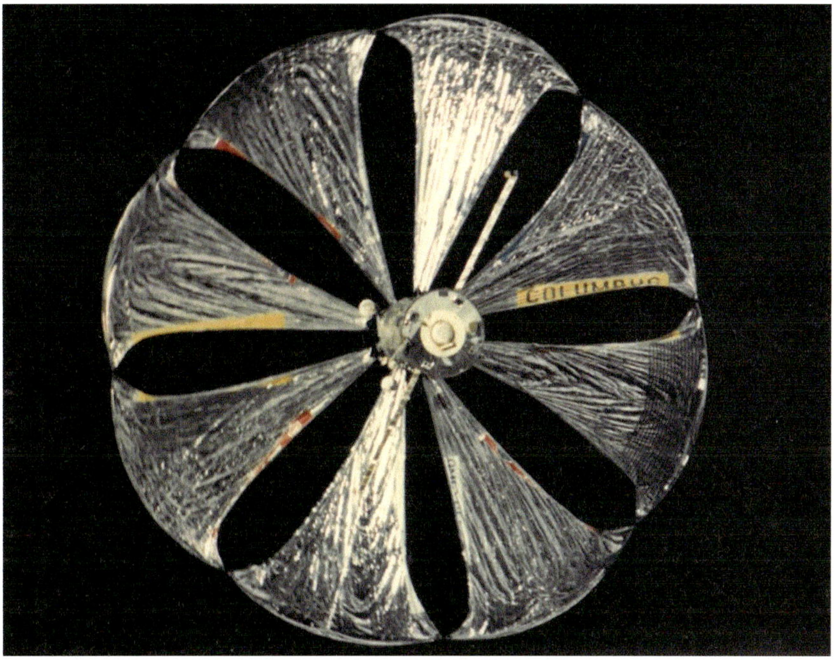

Fig. 3.1 Space experiment *Znamya*-2. Photo from the author's personal archive.

The advantage of the Russian design over its competitors from other countries (recognized by the race organizers) was the absence of any rigid frame holding the film, which reduced the mass launched into orbit and made it easier to deploy the sail. The flat shape of the film was to be maintained by centrifugal forces generated by its rotation. Despite the overall positive results of the *Znamya*-2 experiment, a design flaw was revealed: sectors tensioned radially by centrifugal forces experienced circumferential compressive forces, which caused waves in the film (visible in Figure 3.1), and the sail's reflective properties were worse than expected. This phenomenon is well known in structural mechanics, but for some reason, the designers failed to anticipate its occurrence. The identified flaw was corrected during subsequent research and development (R&D), in which the author was directly involved. An improved *Znamya*-3 project was developed, but after some time its funding was suspended.

Similarly, human factors are "responsible" for gaps and inadequate provisions in the system operating manuals. According to the popular view, "there is no means for quantifying the probability of … inconsistencies in written materials" (Swain and Guttmann, 1983). Thus, the second source of uncertainty, along with the possibility of seeing in it an aleatory nature, obviously has a significant epistemic component.

This component fully characterizes the third cause of uncertainty – the emergence of new properties in the system. Along with the planned useful qualities, after connecting elements into a system (through assembly or other technological operations), a side effect may be obtained in the form of emerging undesirable properties. These properties are sometimes hidden and unexpected, becoming apparent only under particular operating conditions, and for a fundamentally new design they are difficult to predict. An example of emerging undesirable property is the so-called *elastic follow-up* in mechanical joints of structural parts operating at high temperatures and subject to creep. This effect occurs when the stiffness of one of the connected parts differs significantly from that of another. It means the forced following of the more flexible part after the elastic unloading of the more rigid one, which results in increased strains in the former. The effect of such following on the cyclic strength of the connection is a reduction in the predicted service life compared to the value obtained by elastic analysis that does not take this effect into account (RCC-MR, 2007; Jawad and Jetter, 2009; ASME, 2010). After its discovery and explanation of its nature, a significant difference in the stiffness of the connected parts began to be considered as a marker of the possibility of an undesirable property emerging in the structure being created. It is obvious that probabilistic methods for predicting such properties are not applicable. When these properties are insufficiently studied or completely unknown, experimental and theoretical research conducted at the design stage serve to determine them. Presumably, for any of the emerging properties a marker should exist that can indicate at the design stage the manifestation of this property during operation.

Possible uncontrolled impacts and changes in the operating conditions of the system that threaten to exceed the design limits (item 4 in the list above) can be analyzed either within the framework of a probabilistic approach based on statistics, or by postulating such events (outliers) in the form of accidents or emergency situations envisaged in the design. The uncertainty associated with this source lies in the impossibility of determining the time of occurrence of a hazardous event, as well as the difficulty of predicting the intensity of the loads it causes and the behavior of structural elements when limit states are exceeded and subsequent destruction occurs. The traditional design approach provides the ability to manage some of this uncertainty by postulating a hazardous event, but does not cope well with the uncertainty of loads and behavior of the structure under progressive destruction conditions, since the factors of safety are not applicable.

In the aerospace industry, a typical example of a postulated accident is emergency landing, for which an aircraft must be designed – see, e.g., CFR Part 25, § 25.561. In the nuclear power industry, a whole class of *design basis accidents* is provided for (IAEA, 2016; NRC, 2021). To take into account adverse (hazardous) natural events of extreme intensity, statistics of extremes (Gumbel, 1962) or fat-tailed probability distributions (see Sect. 2.2.1) are applicable. The problems with probabilistic analysis of extreme adverse outliers (reflected only in fat tails) are that 1) the length of the statistical series required to ensure the robustness of probability estimates may exceed available experience and 2) when the approximating distribution follows a power law, difficulties arise in limiting the extreme intensity of loads acting on the designed system. When considering these problems, one should also take into account the questions posed in Sect. 2.2.1, in particular the possible causality of outliers by the entry into force of some unknown factors interfering in the aleatory nature of loading processes. Consequently, the fourth source generates uncertainty with a significant epistemic component.

Turning to the source of uncertainty, indicated as item 5 in our list (human factors), we note that probabilistic methods are used to describe the variability of the actions of personnel working with a technical system (Swain and Guttmann, 1983). However, they are poorly suited for analyzing the behavior and safety of other people who come into contact with the technical system during its operation, as well as for assessing the impact on safety of human factors embodied in these people themselves, including (Spirochkin, 2023):

- the public;
- vehicle occupants in crashes;
- attackers (intruders) who have gained unauthorized access to control the system; and
- potential victims of such access.

Probabilistic methods are applicable to the analysis of the behavior and safety of vehicle occupants in crashes, when the processes that characterize behavior and determine safety can be reduced to the field of biophysics or biomechanics due to their high speed. In this case, the safety criteria are expressed in terms of load tolerance and are similar to strength and integrity criteria (ibid). In other manifestations of human factors, consciousness, which controls human behavior, plays such a role that this greatly complicates the analysis. The uncertainty of these manifestations can be classified rather as ambiguity (Sect. 2.3). In any case, this uncertainty has a significant epistemic component. Its analysis is feasible using agent-based mathematical modeling and simulation, but the current level of development of this methodology does not allow solving all safety problems, especially when considering mass human behavior.

With regard to the sixth source of uncertainty – inaccuracies in the experimental and theoretical (computational) substantiations of design decisions – the following considerations can be made. The nature of the incomplete similarity of the material objects used for experimental substantiation (prototypes, research make-ups and small-scale physical models) to the properties of the system being created, as well as the nature of imperfections of mathematical models, errors in computational simulations and software implementing them, is generally not stochastic. It is associated with the impossibility of simultaneously fulfilling all similarity criteria due to fundamental physical laws, limited knowledge and resources, including limitations on the accuracy of numerical operations determined by the architecture of computer systems, etc. At the same time, sources of errors in calculations may be software defects caused by human factors embodied in the code developers. The limited variability inherent in professional activity makes it possible to use a statistically based probabilistic approach to assess the quantitative indicators of such defects. In particular, the capture-recapture method is applicable, which helps to estimate the number of defects remaining in the code and the level of their detection by a row of independent random checks (Schofield, 2007). As a result, the uncertainty corresponding to item 6 of the list of sources can be considered as a combination of a small aleatory part with a predominant epistemic component.

The seventh source of uncertainty in the list we are considering – the possibility of unknown factors – fits entirely into the category of epistemic uncertainty. This possibility should be expected primarily in research and innovation projects involving new or under-developed fields of science and technology. The presence of unknown factors is evidenced by phenomena unexplained by any theory, and facts that contradict each other or do not fit into the modern scientific paradigm. An example of unexplained phenomena is quantum entanglement; among the contradictory facts, it is worth mentioning the observed (judging by the values of redshift and brightness of extremely distant supernovae) accelerated expansion of the Universe[2]. The "unknown unknowns" lurking in the depths of the quantum world or in the vastness of space do not yet have a noticeable impact on the design of any technical systems. Not yet.

Returning to technology on the scale we are accustomed to, we note that unknown factors may be due to the complexity of the system being created, including the complexity of its life cycle processes. Complexity is understood as a specific property of systems in which "large number of components or subsystems, at times of a different nature, combine to produce surprising emergent phenomena apparent at multiple scales" (Hébert-Dufresne et al., 2024). In this context, not only the technical object under development should be considered as a system, but also a set of participants, tools and processes of its creation (production system), as well as a physical complex covering, along with the object, the environment and the processes of interaction with it during operation (operation system). Complexity of a system is often associated with the nonlinearity of its properties and interaction with the environment. Nonlinearity is expressed in the fact that the parameters of the system elements and the connections between them, established for any observed interval of interaction in the time or space domain, change when moving to any other interval, and the previous values become incorrect. As a result, insufficient traceability and controllability of the entire system arises.

Uncertainty caused by the possibility of unknown factors cannot be eliminated completely, but it can be reduced for a certain project using experimental and theoretical methods to obtain new or improved knowledge. Unexplained phenomena and contradictory facts, complexity and nonlinearity should be considered as markers of the action of unknown factors or the possibility of their manifestation in the life cycle of the system.

---

[2]The hypothesis of accelerated expansion of the Universe is not accepted by all cosmologists and is currently being criticized.

3.2 Structuring Uncertainty

The above considerations allow us to imagine a structure of uncertainty in design that serves the purposes of a deeper understanding of this phenomenon and determining priority areas for applying efforts to minimize it. Table 3.1 lists the main sources of uncertainty and shows what components of uncertainty they generate.

Table 3.1 Structured characterization of uncertainty in design.

Source of uncertainty	*Generated components of uncertainty*	
	Aleatory uncertainty	*Epistemic uncertainty*
1. Stochastic variation of the properties and geometry of system elements, as well as maximum loads	Almost entirely	
2. Hidden imperfections of the system and operating documentation	Partially	Partially
3. Emergence of new properties		To the full
4. Uncontrolled impacts on the system and changes in operating conditions	Partially	Partially
5. Human factors	Partially	Partially
6. Inaccuracies in the substantiations of design decisions	As a minor part	As a dominant part
7. Unknown factors		Completely

As can be seen from this table, only one of the seven sources of uncertainty produces almost entirely aleatory uncertainty – the first. The phrase "almost entirely" means that, in addition to the random scatter of mechanical properties and geometry of structural elements, as well as maximum loads, characteristic of each moment in time during the post-design part of the life cycle of a technical system, the trends in the change in their probability characteristics over time should also be taken into account. The means and standard deviations of strength properties may change due to aging of materials under the influence of adverse operational factors, accumulation of damage from fatigue or creep, etc. The geometric parameters of elements are affected by corrosion, erosion and wear. As a result, the mean value of the bearing capacity for a certain statistical population of similar structural elements, corresponding to an expected limit state, $\overline{S}$, decreases and the standard deviation may increase as shown in Figure 3.2. The probabilistic characteristics of the maximum load parameters, $\overline{F}$ and σ_F, inherent in loading modes that can be considered stationary and (or) approximated by thin-tailed distributions, may increase over time – especially in the long term perspective (as shown in the same figure). For loads caused by natural impacts – climatic, meteorological, hydrogeological or seismic, such changes can be determined by global processes in the atmosphere, ocean and lithosphere of the Earth. They may lead to changes in the characteristics currently taken into account in the design using standardized models of natural impacts.

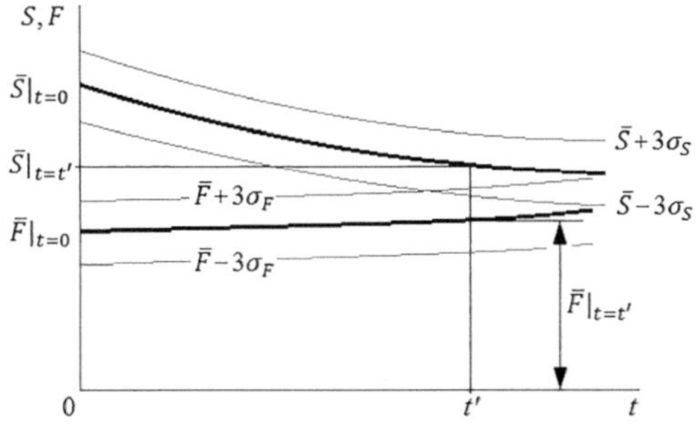

Fig. 3.2 Expected changes in the bearing capacity of structural elements and maximum load parameters during operation of a technical system.

For example, in recent decades, climate change (caused by rising atmospheric carbon dioxide concentration, presumably of anthropogenic origin) has led to an increase in the tropopause height, a poleward shift of the jet streams and their intensification (Lorenz and DeWeaver, 2007), and strengthened clear-air turbulence (Williams and Joshi, 2013; Storer et al., 2017). The probabilistic characteristics of atmospheric turbulence that were derived from extensive studies of this phenomenon in the 1950s and 1960s have begun to transform, currently showing an asymmetric density distribution of the Ellrod turbulence index (Ellrod and Knapp, 1992) with a longer tail indicating a shift to extreme values (Williams and Joshi, 2013). These changes can have a direct impact on the standard atmosphere parameters and maximum aerodynamic loads used in the design of rockets and aerospace vehicles. Commercial airplanes are increasingly encountering extreme turbulence, which results in passenger injuries (sometimes fatal) and damage to the aircraft structure (Rigdon and Paddison, 2025). It is obvious that changing meteorological conditions and the tendency for aerodynamic loads to increase must be taken into account for aircraft in service and even more so for newly developed aerospace vehicles. It is quite likely that similar or other fundamental trends are characteristic of other natural impacts, but the changes that are taking place have not yet been sufficiently studied.

If the random scatter of the parameters and variables under consideration across a given statistical population at any given moment in time falls entirely under the category of aleatory uncertainty, then time trends can carry, albeit to a small degree, uncertainty of a different kind – close to epistemic.

Sources 2, 4, 5, and 6 generate aleatory uncertainty as part of the total uncertainty. The other part is epistemic uncertainty. Methods for reducing aleatory uncertainty and the associated risk mentioned in Sect. 3.1 in relation to the first source (these methods cover quality management at all stages of the life cycle of the system) can be applied to sources 2, 4 and 5 to the extent that such a component manifests itself in them.

Among the problems complicating the probabilistic approach to the aleatory component of uncertainty generated by source 4 (possible uncontrolled impacts on the system and changes in operating conditions), insufficient size of statistical populations and fat tails of approximating distributions were noted in Sect. 3.1. When choosing suitable methods for analyzing such uncertainty, it is necessary to find an answer to the question: can the predicted extreme impacts and abrupt changes in operating conditions be considered stochastic phenomena of the same nature as the similar events previously observed, despite significant differences in intensity? If the answer is yes, then all these phenomena can be combined into one statistical population. However, in this case, the resulting probabilistic characteristics may undergo deformations, since only a small portion of the total population related to extreme events will be replenished. The probability density function of the maximum loads acting on the system will become fat-tailed or positively skewed with a long tail, and this will likely lead to difficulties in obtaining reliable estimates of means, standard deviations, and coefficients of variation. An example is atmospheric turbulence in thunderstorms, which is characterized by extreme intensity compared to clear-air turbulence (Taylor, 1965), as mentioned in Sect. 2.1.2. An acceptable solution would be to consider such extreme manifestations as a separate, additional statistical population, existing alongside the main one. For such an extreme population, probabilistic estimates of the loads determining the static strength of the structure can be more reliable (ibid).

However, the most likely cause of uncontrolled impacts on the system and changes in operating conditions is the entry of new factors – this possibility is noted in Sect. 2.2.1. Identifying such factors and determining the loads and changes they cause, as well as response of the system, goes beyond the stochastic nature of phenomena and the probabilistic approach to them. For this purpose, analytical methods that take cause-and-effect relationships into account are applicable.

The combination of aleatory and epistemic uncertainty in source 5 (human factors) and the diversity of manifestations of these factors, mentioned in Sect. 3.1, necessitate a range of analytical methods – both probabilistic and deterministic. The most promising of these are mathematical modeling and simulation. The presence of epistemic components characterizing some of these manifestations may require fundamentally new approaches to reducing uncertainty.

The specific nature of source 6, which determines the insignificant presence of the aleatory component and the dominance of epistemic uncertainty, also requires methods that go beyond the probabilistic approach. A number of such methods, applicable to the computational substantiations of design decisions, exist and are characterized by a significant degree of elaboration (Oberkampf et al., 2003; Oberkampf and Roy, 2010; Roy and Oberkampf, 2011). However, existing methods do not solve all problems of uncertainty minimization.

As Table 3.1 shows, epistemic uncertainty is fully inherent in sources 3 (emergence of new properties) and 7 (unknown factors). Apparently, to minimize such uncertainty, as well as uncertainty in the form of a combination of aleatory and epistemic components, complex approaches should be used that differ not only from probabilistic ones, but also from those described in Sect. 2.3.

Based on the presented structuring of uncertainty, it is possible to determine a general scheme for its minimization in the design of technical systems.

3.3 Some Non-obvious Issues and Available Tools to Minimize Uncertainty

From a perspective of minimizing uncertainty, the first source listed in Table 3.1 (stochastic variation) would seem to pose fewer problems than the others, since there are well-developed methods for managing the aleatory uncertainty that characterizes it almost entirely. However, the changes in the probabilistic characteristics of the load-bearing capacity and maximum load parameters for a statistical population of similar structural elements over a long life cycle, noted in Sect. 3.2, may become a challenging issue for designers due to the lack of relevant information.

The expected changes include a decrease during operation in the mean of the bearing capacity indicator characterizing a particular limit state, $\overline{S}$ and a possible increase in its standard deviation σ_S (see Figure 3.2). Therefore, for a certain point in time, t', the following expression is valid:

$$\overline{S}|_{t=0} > \overline{S}|_{t=t'}; \ \sigma_S|_{t=0} \leq \sigma_S|_{t=t'}. \tag{3.1}$$

The consequence of this is an increase over time in the corresponding coefficient of variation:

$$w_S|_{t=t'} > w_S|_{t=0}. \tag{3.2}$$

For relevant probabilistic characteristics of the maximum load parameters, $\overline{F}$ and σ_F, in loading modes that appear stationary and (or) are approximated by distributions with thin tails, the worst-case scenario should be considered to be an increase in their values over time:

$$\overline{F}|_{t=t'} > \overline{F}|_{t=0}; \ \sigma_F|_{t=t'} > \sigma_F|_{t=0}, \tag{3.3}$$

as a result of which we obtain:

$$w_F|_{t=t'} \geq w_F|_{t=0}. \tag{3.4}$$

An increase in the coefficients of variation $w_S|_{t=t'}$ and $w_F|_{t=t'}$ leads, according to Eq. (2.9), to an increase in the required value of the factor of safety, f_S, in order to maintain the mechanical reliability and safety of the system at the level specified during design. If it is difficult or impossible to make structural changes to an existing system to strengthen it, the required but not implemented increase in the factor of safety actually means the emergence of a safety deficit during operation. Thus, there is a need to control the stochastic variation of key parameters and variables that determine the mechanical reliability of the system during its life cycle. At the same time, statistics on the loads acting on a population of similar systems during operation should be accumulated and analyzed in order to identify possible deviations that can indicate a change in the initially Gaussian (normal) probabilistic approximations, including a drift towards fat-tailed or asymmetric long-tailed distributions (see Sect. 3.2).

An equally serious challenge may be the introduction in design of new materials, structural configurations and automated control systems, as well as the use of new technologies for the manufacturing of elements and the assembly of the system. The reason is that the application of any innovation, due to the changing random scatter of its characteristics relative to existing prototypes, requires an analysis of this scatter on safety using a representative volume of statistical data. The costs of obtaining this data potentially conflict with the commercial interests of manufacturers and operators of technical systems focused on rapid payback. Among such innovations, which objectively require large-scale research, are advanced composite materials, thin-film space structures, and additive technologies.

The issues described above indicate that, even with a well-developed methodology for probabilistic analysis and the implementation of other risk management measures throughout the life cycle of a technical system, there may be a need to refine the probabilistic characteristics used in design to reduce the aleatory uncertainty associated with source 1. For the reliable function of this entire framework, accurate data is required.

Minimizing the uncertainty created by sources 2, 4, 5, and 6, which represent a combination of aleatory and epistemic components, should obviously be achieved by combining a number of approaches. The aleatory component associated with each of them can be reduced in the same way as for source 1 – by refining the probabilistic characteristics. Along with this, no less important, and perhaps even more relevant, is the identification of approaches suitable for reducing the epistemic component of uncertainty. Such approaches are more in demand for sources 3 and 7.

However, any attempt to manage epistemic uncertainty inevitably raises the question: how can we assess the reduction in epistemic uncertainty, which is assumed to be immeasurable? Based on the characterization of such uncertainty given in Sect. 2.1.1, which associates it with the absence of knowledge about all possible outcomes of an event in question or all possible events in a given context, the following criterion of reduction can be proposed: epistemic uncertainty decreases whenever knowledge is acquired about some outcome or event that was previously absent. A distinctive feature of this criterion is its relative nature, not tied to specific numerical values. By adopting this criterion, we will consider the minimization of epistemic uncertainty as a process of acquiring new knowledge it can be viewed as expanding the boundaries of our knowledge into the previously unknown.

Thus, minimizing uncertainty in the design of a technical system should include refining the values of known variables and parameters characterizing its life cycle, under conditions of inevitable stochastic variation – up to the possibility of determining these values for each specimen of the system at each moment in time, as well as obtaining knowledge about previously unknown properties and phenomena.

Engineering tools applicable to uncertainty minimization include analytical methods, experiments, new design concepts, and technologies implemented at different stages of the life cycle. These can be used in various combinations depending on the sources of uncertainty. The purposeful and justified application of any single tool or a combination of several tools constitutes an engineering approach to addressing uncertainty. A number of such approaches are described in the following chapters of this book. Some of them have long been used in design practice, and the purpose of their description is to systematize them, clarify the scope of application, and identify their shortcomings. Other, more modern and sophisticated approaches are being implemented and tested in advanced industries. Along with well-known methods for reducing uncertainty, new concepts currently under development are presented below.

References

ASME (2010) An International Code. 2010 ASME Boiler & Pressure Vessel Code. Section III. Rules for construction of nuclear facility components. Division 1, Subsection NH – Class 1 Components in elevated temperature service. American Society of Mechanical Engineers (ASME), New York.

CFR Part 25. Code of Federal Regulations/Title 14/Chapter I/Subchapter C/Part 25 – Airworthiness Standards: Transport Category Airplanes. https://www.ecfr.gov/current/title-14/chapter-I/subchapter-C/part-25. Accessed July 11, 2025.

Ellrod, G.P. and Knapp, D.L. (1992) An objective clear-air turbulence forecasting technique: Verification and operational use. *Weather Forecasting*, Vol. 7, pp. 150–165.

Gladkii, V.F. (1982) Veroyatnostnye metody proektirovaniya konstruktsii letatelnogo apparata *(Probabilistic methods for designing spacecraft structures)*. Nauka, Moscow (in Russian).

Gumbel, E. (1962) *Statistics of extremes*. Columbia University Press, New York.

Hébert-Dufresne, L., Allard, A., Garland, J., Hobson, E.A. and Zaman, L. (2024) *npj Complexity*, 1(4). https://doi.org/10.1038/s44260-024-00004-0.

IAEA (2016) IAEA Safety Standards Series No. SSR-2/1 (Rev.1). Safety of nuclear power plants: Design. Specific safety requirements. International Atomic Energy Agency, Vienna.

Jawad, M.H., Jetter, R.I. (2009) *Design and analysis of ASME boiler and pressure vessels components in the creep range*. ASME Press, New York.

Lorenz, D.J., and DeWeaver, E.T. (2007) Tropopause height and zonal wind response to global warming in the IPCC scenario integrations. *Journal of Geophysical Research*, 112, D10119.

McEvily, A.J. (2002) *Metal failures: mechanisms, analysis, prevention*. John Wiley & Sons, New York.

NRC (2021) Design basis accident. U.S. NRC – United States Nuclear Regulatory Commission. March 09, 2021. https://www.nrc.gov/reading-rm/basic-ref/glossary/design-basis-accident. Accessed July 27, 2025.

Oberkampf, W.L., Trucano, T.G., and Hirsch, C. (2003) Verification, validation, and predictive capability in computational engineering and physics. Report SAND2003-3769. Sandia National Laboratories, Albuquerque, NM, and Livermore, CA.

Oberkampf, W.L., Roy, C.J. (2010) *Verification and validation in scientific computing*. Cambridge University Press, Cambridge, UK.

RCC-MR (2007) *Design and construction rules for mechanical components of nuclear installations*. Section 1 – Nuclear installation components. AFSEN, Paris.

Rigdon, R., Paddison, L. (2025) Turbulence is getting worse. Here are the planet's bumpiest flight routes. *CNN Climate*, August 07, 2025. https://edition.cnn.com/2025/08/07/climate/flight-turbulence-worst-routes-vis. Accessed August 08, 2025.

Roy, C.J., Oberkampf, W.L. (2011) A comprehensive framework for verification, validation, and uncertainty quantification in scientific computing. *Computer Methods in Applied Mechanics and Engineering*, Vol. 200, pp. 2131–2144.

Schofield, J. (2007) Beyond defect removal: latent defect estimation with Capture-Recapture Method. *CrossTalk, The Journal of Defense Software Engineering*, 20(8):27–29.

Spirochkin, Y. (2023) *Human factors and design*. Springer, Singapore.

Spirochkin, Y.K., Saprykin, O.A. (2021) Kosmicheskii polyot pod solnechnym parusom (Space flight under a solar sail). *AviaSafety.ru*, January 6, 2021 (in Russian). https://aviasafety.ru/36409. Accessed November 23, 2025.

Storer, L.N., Williams, P.D., and Joshi, M.M. (2017) Global response of clear-air turbulence to climate change. *Geophysical Research Letters*, 44.

Sukhoruchenkov, B.I., Okorokov, M.V. (2014) Estimates of indicators of possible defects of technical systems not identified after development tests. *Dual technologies*, No. 1(66), pp. 12-18 (in Russian).

Swain, A.D., Guttmann, H.E. (1983) Handbook of human reliability analysis with emphasis on nuclear power plant applications. NUREG/CR-1278. SAND80-0200. RX, AN. Final report. Sandia National Laboratories, Albuquerque, NM.

Taylor, J. (1965) *Manual on aircraft loads*. Pergamon Press, Oxford; London.

Williams, P.D. and Joshi, M.M. (2013) Intensification of winter transatlantic aviation turbulence in response to climate change. *Nature Climate Change*, 3, pp. 644–648. https://doi.org/10.1038/nclimate1866.

4 Reducing Uncertainty by Analytical Methods

This chapter considers approaches to uncertainty reduction based on analytical methods. Uncertainty that can be reduced in such a way corresponds to inaccuracies in the experimental or theoretical (computational) substantiations of design decisions and is predominantly epistemic. Among the tools available to designers for reducing this uncertainty, analytical methods are obviously the least expensive. They rely on a well-developed theoretical foundation. Lower costs and a sufficient level of development determine the priority given to these methods. The consideration covers widening the object of analysis, refining the mathematical models used in the analysis, reconstructing the phenomenon being analyzed based on available fragments of information, and a number of other analytical tools. The chapter ended with the general assessment of the applicability of the analytical approach to uncertainty reduction.

4.1 Widening the Object of Analysis

4.1.1 Application Area: Coupled Problems

The method of widening the object of analysis is often necessary when substantiating the strength of a technical system being created in the design cases characterized by the interaction of its structural elements with the environment. As a rule, strength substantiation is performed by calculating the displacements, strains and stresses in structural elements under the action of applied loads and their assessment using strength criteria. Load values are typically determined in aerodynamic, thermal, hydraulic, or other calculations that precede the strength analysis and provide data on the state and behavior of the environment (both external and internal) with which the structure interacts. However, it is not always possible to determine the behavior of the environment independently of the mechanical response of structural elements, since this behavior and the loads caused by it may depend on the response. Problems in which the loading process depends on the structural response are categorized as *coupled*. If this dependence and the coupled nature of the interaction between the structure and the environment are not taken into account, then uncertainties (errors) arise in the values of the structural response parameters obtained in a separate calculation.

To demonstrate the method of widening the object of analysis, we consider a number of examples. Before considering these, we note that the role of the environment can be played by a liquid or gas with which structural elements interact,

DOI: 10.1201/9781003670575-4

as well as other parts of the system to which these elements are mechanically connected but which are of no interest to the current analysis.

4.1.2 Connection of a Structural Element with the Main Part of the System

The first example concerns the case where the environment is formed by a structural connection between the element in question (*A*) and a set of other elements that comprise the main part (*B*) of the designed technical system, as shown in Figure 4.1. This connection is made using elastic attachment units, and the entire system is subject to dynamic load *F*(*t*) applied to its main part. The loading of the element occurs through kinematic excitation of its vibrations at the attachment units.

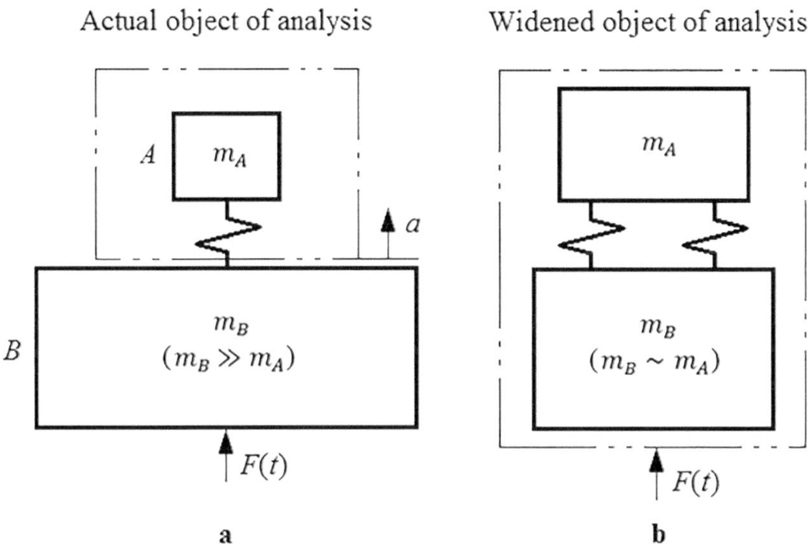

Fig. 4.1 Widening the object of analysis in the case of a structural connection between this object (*A*) and the main part of the system (*B*).

Kinematic loading cases are typical for cargo carried by aircraft when hitting atmospheric turbulence or when landing, as well as for NPP equipment and pipelines during an earthquake. Loads acting on attached elements are specified for the design by means of acceleration *a* in the form of:

- its maximum value at the center of gravity of the element, a_{max};
- the dependence of acceleration on time (accelerogram), $a(t)$;
- the dependence of maximum acceleration on the frequency of vibrations, f, and the damping ratio, ξ, i.e. response spectrum, or shock spectrum, $a_{max}(f, \xi)$.

However, this method of specifying loads is applicable only in cases where it is possible to neglect the influence of vibrations of the element on the structural dynamics of the main part of the system. And this influence can be considered negligibly small if the mass of the element, m_A, is significantly less (at least an order of value) than the mass of the main part of the system to which it is connected, m_B. Then the object of strength analysis can only be the element itself, and the analysis can be carried out by calculating its mechanical response to kinematic excitation, specified by acceleration a (see Figure 4.1, image a).

Otherwise, if the element is comparable in mass to the main part of the system: i.e. $m_A \sim m_B$, then its vibrations can have a significant effect on the dynamic behavior of this part. If this effect is not taken into account, it gives rise to epistemic uncertainty in the form of errors in determining the dynamic response of the element and substantiating its strength (as well as the strength of the attachment units). Such uncertainty can not only be reduced, but eliminated completely by widening the object of analysis – by including the main part of the system in the calculation (Figure 4.1, image b). Since strength analysis is usually carried out using a computational mathematical model, widening the object of analysis means practically the extension of the modeling beyond the element to the main part of the structure. The widened analysis is performed based on the load acting on the main part of the system, $F(t)$. Of course, the question remains as to how substantiated this load itself is, but its determination in this case goes beyond the scope of strength analysis and refers to other calculations or experiments preceding this analysis.

4.1.3 Vibration of a Structure in Contact with a Liquid

As a second example, we consider the vibrations of a structure that contains liquid or (and) is immersed in it. The first case is typical for launch vehicles, the structure of which includes large fuel and oxidizer tanks, and the second is for nuclear reactors of the WWER type, which have a pressure vessel inside which a shaft with the reactor core is located and the coolant – water – circulates, as shown in Figure 4.2.

When internal or external time-varying loads excite vibrations in the reactor structure, vibrations are also initiated in the liquid, creating hydrodynamic forces that change the dynamic response spectrum of the structure compared to the case of no liquid (a "dry" structure). The changes consist of a decrease in the vibration frequencies due to the effect of the added mass of the liquid and the appearance of new vibration modes corresponding to the movement of the layers of liquid adjacent to its free surface, with the formation of waves on it. Such interaction of the structure with the liquid determines the need to include the volumes it occupies in the analysis of the dynamic response of the structure. In this case, the actual object of analysis (structure) is widened to the mechanical system "structure-liquid", and accordingly the boundaries of its mathematical modeling are expanded. This method is used in the dynamic analysis of the behavior of launch vehicles (Abramson, 1966; Mikishev and Rabinovich, 1971; Dodge, 2000), in determining the dynamic characteristics of nuclear reactor structures (Spirochkin, 1997, 1998), and in a wide range of similar problems requiring consideration of liquid motion

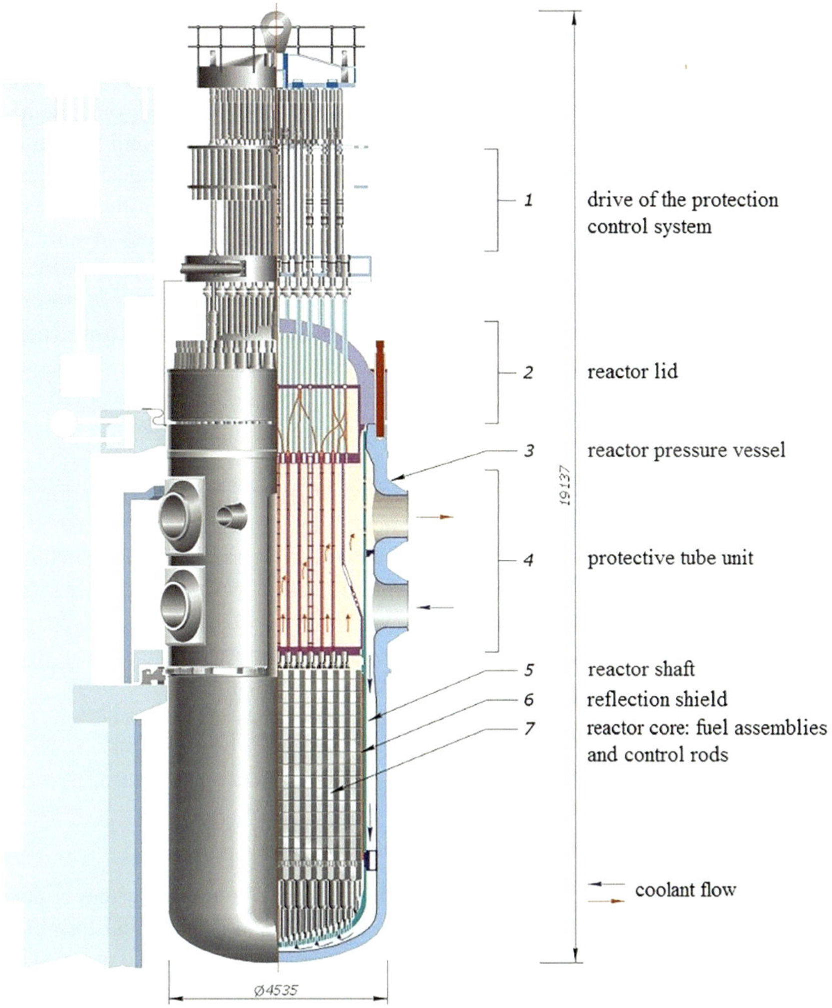

Fig. 4.2 Nuclear reactor WWER-1000. Reproduced from the image by Panther. https://upload.wikimedia.org/wikipedia/commons/6/60/Wwer-1000-scheme.png. CC BY-SA 3.0.

(sloshing) in reservoirs (Ibrahim, 2005). The uncertainty of the results of solving each of these problems is reduced to a level determined by the accuracy of input data and the perfection of the mathematical model used (see Sect. 4.2).

4.1.4 Aero- and Hydroelasticity

The third example refers to the phenomena of aero- and hydroelasticity that arise during the relative movement of a flexible structure and a fluid at a certain velocity. These cover various types of loss of stability of elastic structural elements and are typical for aircraft flying in the atmosphere, pipelines of energy facilities when a water coolant or a steam-air mixture moves through them, etc. In aerospace engineering, such phenomena primarily include *divergence* (static loss of stability) and *flutter* (vibration with an uncontrollably increasing amplitude, or dynamic loss of stability). Loss of stability at some previously unknown combination of mechanical characteristics of structural elements, fluid density and kinematic parameters of the flow represents a potential hazard, since an uncontrolled increase in displacements can lead to collapse of the structure. Therefore, loss of stability must be eliminated at the design stage, which requires determining the conditions for its occurrence through appropriate calculations.

In this case, the actual (of interest to designers) object of analysis is a structural element in question or a set of mechanically connected elements that may be subject to divergence or flutter, but the analysis must also include volumes of fluid that enter into aero- or hydrodynamic interaction with them. As in the previous example, such widening the object of analysis provides a reduction in uncertainty regarding the occurrence of aero- or hydrodynamic instability to a level corresponding to the accuracy of the data on the characteristics of the structure and the quality of the mathematical description of the interaction between the structure and the fluid.

Analysis using mathematical models and calculation methods adequate to the characteristics of the structure and fluid flow (Volmir, 1967, 1976, 1979; Wright and Cooper, 2007; Dowell, 2015) allows designers to determine the critical flow velocity at which loss of stability occurs and to identify structural behavioral features. As a result, measures can be taken to control aero- or hydroelasticity.

4.2 Refinement of Mathematical Models

4.2.1 Mathematical Models and Related Errors

A mathematical model represents an object of analysis (a technical system or part of it) using various mathematical entities: geometric figures, analytical formulas or computational algorithms to determine its behavior, i.e., response to external or internal impacts (loads). The determination of the response is based on quantitative data about these loads and the properties of the system (characterizing its ability to respond in the manner under consideration). In general, such a determination can be viewed as a procedure for solving an operator equation:

$$\mathbb{R} = \mathfrak{M} \times \mathbb{F}. \tag{4.1}$$

The left side of this equation, the set $\mathbb{R}$, contains variables that characterize the response to be determined. The member $\mathfrak{M}$ is a *model operator* that includes mathematical operations that are performed to obtain these variables in an explicit quantitative form. The set $\mathbb{F}$ contains input data on loads.

It should be noted that the term "mathematical model" is vague, as it encompasses very different concepts, including:

- a set of geometric primitives that reflects the configuration of the object of analysis (geometric model);
- a system of the governing algebraic, differential or integral equations that describe the analyzed physical process (analytical model);
- an approximation of these equations within the framework of the applied numerical method (numerical model);
- the implementation of the numerical model in the form of an algorithm, as well as its embodiment in computer software (computational model).

The model operator covers all these options, processing the input data, $\mathbb{F}$, and producing the output, $\mathbb{R}$. Thus, it is a type of mathematical model. A mathematical model is constructed based on the nature of loads being considered and the need to determine the object's response to them, reflecting the basic characteristics of the object and the essential features of its response that are of interest. To construct the model, an appropriate modeling method is employed, along with certain simplifying assumptions that allow Eq. (4.1) to be solved using analytical tools and computational resources available to the analysis engineer or researcher.

The input data on loads, regardless of whether they were obtained experimentally or by calculation, inevitably contain uncertainty in the form of inaccuracies (errors). We denote these errors by the set $\mathbb{E}_{\mathbb{F}}$. As a result of processing these errors as a component of the set $\mathbb{F}$ by the model operator in accordance with Eq. (4.1), they are transformed into response uncertainty – as a set of the corresponding errors $\mathbb{E}_{\mathbb{R}}$. This transformation is affected by the uncertainty associated with errors in the mathematical model itself.

In order to identify sources of errors that may arise in mathematical models, including errors in the model operator, let us turn to a specific application area – the strength analysis of a structure under design. Currently, such analyses are almost always carried out using mathematical models constructed by the *finite element method* (FEM). It came into use more than half a century ago (Zienkiewicz, 1971) and is now the dominant numerical method for modeling structures and continuous media. Descriptions of this method can be found in numerous publications. Below is a brief introduction to FEM, essential for understanding the ideas that follow. It should be noted that we will not discuss the errors that analysis engineers or researchers may make when using this method incorrectly, but rather focus on the errors inherent in the method itself.

4.2.2 Brief Introduction to the Finite Element Method as a Method of Mathematical Modeling

To analyze the mechanical behavior of a structure or a continuous medium using FEM, a geometric model of the object of analysis is constructed in the form of a spatial mesh – an ensemble of finite elements, as shown in Figure 4.3.

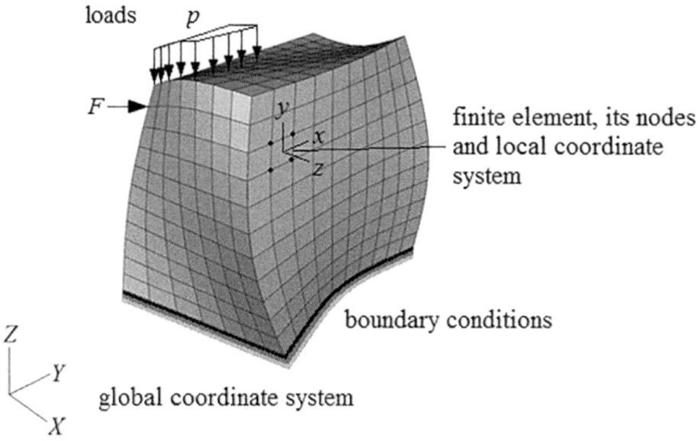

Fig. 4.3 Finite element model.

In the analysis of the strength of a structure (as well as a solid medium), the response variables are typically displacements, strains and stresses at the points of the object of analysis. Displacements are the main variables to be determined, and strains and stresses are calculated from them using the relations of elasticity or plasticity theories. Displacements at any point of a finite element with coordinates x, y, z in its local coordinate system, $\{u\}$, are expressed through displacements at the mesh nodes belonging to this element, $\{u\}_e$, using interpolation functions $N(x, y, z)$, or *shape functions*, determined within the finite element:

$$\{u\} = \{u(x, y, z)\} = [N(x, y, z)]\{u\}_e, \tag{4.2}$$

where $\{u(x, y, z)\} = [u_x \ u_y \ u_z \ ...]^T$ (in addition to linear displacements along the axes x, y, z, also rotation angles and other kinematic parameters can be considered);

$\{u\}_e = [u_{1x} \ u_{1y} \ u_{1z} \ ... \ u_{2x} \ u_{2y} \ u_{2z} \ ...]^T$ is the column vector of nodal displacements of the finite element (numeric indices correspond to the numbers of its nodes);

$[N(x, y, z)]$, or simply $[N]$, is the matrix of shape functions.

Finite elements can have different spatial dimensionalities and geometrical forms depending on the configuration of the modeled object and the problem being solved, as well as different number of nodes and variables considered at each node – degrees of freedom. Polynomials are most often used as shape functions. The order and number of terms of such polynomials, characterizing, along with spatial configuration, the type of finite element, are chosen based on the analyzed stress-strain state. This choice is also influenced by the number of nodal degrees of freedom of the element, since the coefficients before the polynomial terms are determined through the values of the response variables at the nodes.

For example, to analyze uniaxial tension-compression of truss structural elements, rod finite elements with two nodes are used, and in each node one degree of freedom is considered – axial displacement. The corresponding displacements at the internal points of such finite element are described by a first-order polynomial:

$$u_x = a_0 + a_1 x. \tag{4.3}$$

The coefficients a_0 and a_1 are determined based on the conditions:

$$u_x|_{x=0} = u_{1x};\ u_x|_{x=l} = u_{2x}, \tag{4.4}$$

where l is the length of the finite element.

As a result, Eq. (4.3) takes the form:

$$u_x = \begin{bmatrix} N_{1x} & N_{2x} \end{bmatrix} \begin{Bmatrix} u_{1x} \\ u_{2x} \end{Bmatrix} \begin{bmatrix} 1 - \frac{x}{l} & \frac{x}{l} \end{bmatrix} \begin{Bmatrix} u_{1x} \\ u_{2x} \end{Bmatrix}. \tag{4.5}$$

To analyze the bending of extended structural elements, two-node beam finite elements can be used. In each node of the beam finite element, two pairs of degrees of freedom are considered to describe the bending in each of the two directions: the lateral displacement, u_y, and the angle of rotation of the cross sections around the z axis, φ_z, as well as the side displacement, u_z, and the corresponding angle, φ_y. The lateral displacements are approximated by third-order polynomials:

$$u_y = [N_{1y}\ \ N_{1\varphi}\ \ N_{2y}\ \ N_{2\varphi}] \begin{Bmatrix} u_{1y} \\ \varphi_{1z} \\ u_{2y} \\ \varphi_{2z} \end{Bmatrix};$$

$$N_{1y} = 1 - 3\left(\frac{x}{l}\right)^2 + 2\left(\frac{x}{l}\right)^3;\ N_{1\varphi} = x\left(1 - \frac{x}{l}\right)^2; \tag{4.6}$$

$$N_{2y} = 3\left(\frac{x}{l}\right)^2 - 2\left(\frac{x}{l}\right)^3;\ N_{2\varphi} = \frac{x^2}{l}\left(\frac{x}{l} - 1\right),$$

where φ_{1z} and φ_{2z} are the angles of rotation of the cross sections in the first and second nodes of the finite element, respectively. Similar relationships are used for the side displacements.

To model structural elements with a flat stress-strain state, such as membranes, finite elements in the form of triangles and quadrangles are used, the displacements of the points of which are described by polynomials of two variables: x and y. For a triangular finite element with three nodes, these polynomials are of the first order:

$$\begin{aligned} u_x &= a_0 + a_1 x + a_2; \\ u_y &= a_3 + a_4 x + a_5 y. \end{aligned} \tag{4.7}$$

The membrane displacements of a rectangular finite element with four nodes are described by second-order polynomials:

$$u_x = a_0 + a_1 x + a_2 y + a_3 xy;$$
$$u_y = a_4 + a_5 x + a_6 y + a_7 xy. \tag{4.8}$$

Modeling of structural elements such as plates and shells subject to bending is carried out using plate (shell) finite elements. The simplest of them are triangular and rectangular elements with three and four nodes, respectively. In each node, three degrees of freedom are considered: displacement along the normal to the element surface, u_z, and two angles of rotation: $\varphi_x = -\partial u_z/\partial y$ and $\varphi_y = \partial u_z/\partial x$ (in this case the local coordinate system of the finite element is left-handed).

The bending displacements of the simplest rectangular finite element are approximated by a fourth-order polynomial (Zienkiewicz, 1971):

$$u_z = a_0 + a_1 x + a_2 y + a_3 x^2 + a_4 xy + a_5 y^2 + a_6 x^3 + a_7 x^2 y + a_8 xy^2 + a_9 x^3 + a_{10} x^3 y + a_{11} xy^3. \tag{4.9}$$

When analyzing the three-dimensional stress-strain state of massive structural elements or a solid medium, solid finite elements in the form of tetrahedrons and hexahedrons are used for their modeling. At each node of such elements, three degrees of freedom are considered – displacements u_x, u_y, u_z. For the simplest tetrahedron with three nodes, each of these displacements is approximated by a first-order polynomial, similar to Eq. (4.7), for example:

$$u_x = a_0 + a_1 x + a_2 y + a_3 z, \text{ etc.} \tag{4.10}$$

In all the listed cases, the coefficients before the polynomial terms and, accordingly, the resulting shape functions, are determined through the displacements at the nodes of the finite element, which play the role of local boundary conditions, as demonstrated by Eq. (4.4). When using polynomials as shape functions, the accuracy of the approximation of response variables can be improved by increasing their order. For this purpose, additional nodes are inserted on the edges of two-dimensional finite elements or on the facets of three-dimensional solid elements. The use of isoparametric finite elements, in which the same additional nodes are used to determine their geometry and the response variable (Zienkiewicz, 1971), also contributes to a more accurate approximation. Such finite elements can be conformal, i.e. they can more accurately model curvilinear elements of the structure.

Based on the chosen shape functions, matrices of finite elements are determined that reflect the mechanical properties of the modeled object. One of them is the *stiffness matrix*:

$$[K]_e = \int_V [B]^{\mathrm{T}}[D][B]dV, \tag{4.11}$$

where $[B]$ is the matrix of strain functions that relates the strains at a point of the finite element, $\{\varepsilon\}$, to its nodal displacements $\{u\}_e$:

$$\{\varepsilon\} = [B]\{u\}_e; \tag{4.12}$$

$[D]$ is the elasticity matrix of the material, which is used to determine stresses, $\{\sigma\}$:

$$\{\sigma\} = [D]\{\varepsilon\}; \tag{4.13}$$

V denotes the finite element volume.

For example, the analysis of a three-dimensional stress-strain state involves considering three tensile or compressive strains, ε_x, ε_y, ε_z, and three shear strains, γ_{xy}, γ_{yz}, γ_{zx}:

$$\{\varepsilon\} = \begin{bmatrix} \varepsilon_x & \varepsilon_y & \varepsilon_z & \gamma_{xy} & \gamma_{yz} & \gamma_{zx} \end{bmatrix}^{\mathrm{T}}. \tag{4.14}$$

These strains are determined as follows:

$$\begin{gathered} \varepsilon_x = \frac{\partial u_x}{\partial x};\ \varepsilon_y = \frac{\partial u_y}{\partial y};\ \varepsilon_z = \frac{\partial u_z}{\partial z}; \\ \gamma_{xy} = \frac{\partial u_x}{\partial y} + \frac{\partial u_y}{\partial x};\ \gamma_{yz} = \frac{\partial u_y}{\partial z} + \frac{\partial u_z}{\partial y};\ \gamma_{zx} = \frac{\partial u_z}{\partial x} + \frac{\partial u_x}{\partial z}. \end{gathered} \tag{4.15}$$

Since the displacements u_x, u_y, u_z are expressed through shape functions according to Eq. (4.2), it is obvious that the matrix $[B]$ contains the partial derivatives of the corresponding functions $N(x, y, z)$ on local coordinates. The strains under consideration cause normal and shear stresses in the material:

$$\{\sigma\} = \begin{bmatrix} \sigma_x & \sigma_y & \sigma_z & \tau_{xy} & \tau_{yz} & \tau_{zx} \end{bmatrix}^{\mathrm{T}}. \tag{4.16}$$

In the case of an isotropic material, the elasticity matrix can be represented as follows:

$$[D] = \frac{E(1-\mu)}{(1+\mu)(1-2\mu)} = \begin{bmatrix} 1 & \frac{\mu}{1-\mu} & \frac{\mu}{1-\mu} & 0 & 0 & 0 \\ & 1 & \frac{\mu}{1-\mu} & 0 & 0 & 0 \\ & & 1 & 0 & 0 & 0 \\ & & & \frac{1-2\mu}{2(1-\mu)} & 0 & 0 \\ & \text{symmetrically} & & & \frac{1-2\mu}{2(1-\mu)} & 0 \\ & & & & & \frac{1-2\mu}{2(1-\mu)} \end{bmatrix}, \tag{4.17}$$

where E is Young's modulus, and μ is Poisson's ratio.

For simple finite elements in the form of linear segments, triangles, rectangles, tetrahedrons or hexahedrons, all nodes of which coincide with the end points or vertices of these geometric entities, under the assumption of constant elastic properties over the volume of the element, the matrices $[K]_e$ are obtained according to Eq. (4.11) in the form of analytical expressions (formulas). For finite elements of complex geometry, including isoparametric ones, the matrices $[N]$ and $[B]$ are determined in curvilinear coordinates oriented differently at different points of the element. Their stiffness matrices are calculated by numerical integration using quadrature points and Jacobian matrices to transform variables from curvilinear coordinates to Cartesian ones (Zienkiewicz, 1971).

By summing up the stiffness matrices of the finite elements according to their connections between themselves and spatial positions, we obtain the stiffness matrix of the finite element ensemble corresponding to the object of analysis as a whole:

$$[K] = [k_{ij}] = \sum [T]_e^{\mathrm{T}} [K]_e [T]_e \qquad (i, j = 1, 2, \ldots n_{dof}), \tag{4.18}$$

where $[T]_e$ is the transformation matrix from global coordinates X, Y, Z to local coordinates of the finite element, x, y, z;

n_{dof} is the number of degrees of freedom of the ensemble.

The numerical model constructed within the framework of the described methodology, in the case of its application to the strength analysis of a structure loaded with static forces, under the assumption of small displacements and the absence of plastic strains, can be expressed as a system of linear algebraic equations:

$$[K]\{U\} = \{F\}, \tag{4.19}$$

where $\{U\}$ is the column vector of nodal displacements of the finite element ensemble, which are the response variables of interest (determined at the mesh nodes), and

$\{F\}$ is the column vector of nodal forces equivalent to the applied load.

The determination of $\{F\}$ is easy when the load consists of forces applied at the points coinciding with the nodes of the finite element mesh. If the load is distributed over the surface of the structure, and for each loaded finite element in its local coordinates can be specified as $\{p\} = [p_x \quad p_y \quad p_z \quad ...]^{\mathrm{T}}$ (p_z corresponds to normal pressure; p_x and p_y are tangential stresses), then the column vector of equivalent nodal forces is calculated for this element:

$$\{F\}_e = \int_A [N]^{\mathrm{T}} \{p\} dA, \tag{4.20}$$

where A is the area over which the load is distributed within the element.

For simple finite elements (see above), the components of the column vector $\{F\}_e$, like the components of the stiffness matrix, are determined according to this equation in analytical form. For finite elements of complex geometry, integration is performed numerically using quadrature points and Jacobian matrices.

The obtained members of $\{F\}_e$ are included in the column vector of equivalent nodal forces of the finite element ensemble:

$$\{F\} = \sum [T]_e^{\mathrm{T}} \{F\}_e. \tag{4.21}$$

The construction of a finite element model (a numerical model in our classification) is completed by including in Eq. (4.19) the boundary conditions characteristic of the problem being solved.

If, when substantiating the strength of the designed structure, an analysis of its dynamic response is required, i.e. the calculation of vibrations caused by time-varying loads, $\{F(t)\}$, then the finite element model, constructed under the assumption of small elastic vibrations, without taking into account the propagation of deformation waves, has the form of a system of ordinary differential equations:

$$[M]\{\ddot{U}\} + [C]\{\dot{U}\} + [K]\{U\} = \{F(t)\}, \tag{4.22}$$

where $[M]$ is the *inertia matrix* of the finite element ensemble;

$\{\ddot{U}\}$ contains the accelerations at the nodes of this ensemble: $\{\ddot{U}\} = \dfrac{d^2}{dt^2}\{U\}$;

$\{C\}$ is the *damping matrix* of the finite element ensemble that describes the dissipation of energy during vibrations due to internal viscous friction in the structure or solid medium;

$\{\dot{U}\}$ contains the nodal velocities of the ensemble: $\{\dot{U}\} = \dfrac{d}{dt}\{U\}$.

The inertia matrix is assembled from the inertia matrices of individual finite elements, similar to the formation of the stiffness matrix of the ensemble:

$$[M]=[m_{ij}]=\sum[T]_e^{\mathrm{T}}[M]_e[T]_e \quad (i,\, j=1,2,\ldots n_{dof}). \tag{4.23}$$

The inertia matrix of a finite element is determined as follows:

$$[M]_e=\int_V [N]^{\mathrm{T}}[\rho][N]dV, \tag{4.24}$$

where $[\rho]$ is the diagonal matrix of distributed inertia properties of the element, relating accelerations and inertial forces and moments, referred to a unit of volume. The components of this matrix for linear displacements are equal to the density of the material, ρ. In the case of taking into account the angles of rotation, the matrix $[\rho]$ includes moments of inertia of the finite element cross sections. For example, for plate (shell) elements, the components $\rho\frac{h^2}{12}$ are used, where h is the thickness of the plate or shell.

For simple finite elements (see above), the inertia matrices are determined in analytical form, similar to the determination of $[K]_e$. If the finite element has complex geometry, numerical integration is used to obtain the inertia matrix.

The damping matrix $[C]$ can be obtained in the same way – by assembling the corresponding matrices of the individual finite elements. When the damping properties for individual finite elements are not determined, but it can be assumed that the nature of internal friction is uniform throughout the structure, it is acceptable to represent the matrix $[C]$ as a linear combination of the inertia and stiffness matrices of the ensemble (Bathe and Wilson, 1976):

$$[C]=a[M]+b[K], \tag{4.25}$$

where a and b are coefficients calculated from reference data, including recommended values of relative damping.

To complete the construction of the dynamic finite element model, the initial conditions are added to Eq. (4.22):

$$\text{at } t=0 \quad \{U\}=\{U\}_0,\ \{\dot{U}\}=\{\dot{U}\}_0, \tag{4.26}$$

where $\{U\}_0$, $\{\dot{U}\}_0$ are nodal displacements and nodal velocities of the ensemble at the initial moment of time. These column vectors take into account the boundary conditions.

FEM allows for the construction of numerical models beyond the simplifying assumptions under which Eqs. (4.19) and (4.22) were obtained, particularly for cases of large displacements, plastic strains, and other manifestations of nonlinear behavior of a structure or solid medium. These models will have a different form than the equations mentioned, but the modeling methodology remains essentially the same as described above. FEM can be used to analyze other physical processes beyond structural mechanics, such as heat transfer, liquid or gas flow, electromagnetic phenomena, and so on. Such models can be applied to calculate the loads acting on the structure. FEM is also applicable to modeling the coupled behavior of the structure and its environment, thus enabling the implementation of the approaches to widening the object of analysis described in Sect. 4.1.

Equations (4.19) and (4.22), as well as similar equations obtained in various applications of FEM, do not correspond to the operator form characteristic of Eq. (4.1). Methods for transforming them to this form pertain not to FEM itself, but to the numerical solution of FEM equations. These methods are presented below, in Sect. 4.2.3, where the errors in finite element models and in the responses calculated using them are discussed.

4.2.3 Errors in Finite Element Models and in the Responses Determined Using Them

As follows from the above brief introduction to FEM, the main causes of errors in finite element models are:

1) limitation of the set of considered response variables inside a finite element, $\{u\}$, and at its nodes, $\{u\}_e$ – due to this some features of the behavior of the structure under load are not taken into account (for example, when describing the bending of plates and shells only using variables u_z, φ_x and φ_y, which is typical for simple triangular and quadrangular elements with nodes at the corners, shear strains of the cross sections are neglected);
2) approximation of response variables by the shape functions $[N]$ that do not accurately reflect the solutions of the differential equations describing the behavior of the structure (the equations of elasticity theory);
3) approximation of parts of a structure using non-conformal finite elements (which do not accurately represent the geometry of these parts);
4) change in the properties of material over the volume of a finite element, leading to inaccuracies in their average values used in the matrices $[D]$ and $[\rho]$;
5) determination of the stiffness or inertia matrices for complex finite elements by the numerical integration procedure according to Eq. (4.11) or Eq. (4.24), using quadrature points and Jacobian matrices;
6) determination of nodal forces equivalent to distributed loads according to Eq. (4.20) using the same numerical integration procedure;
7) representation of boundary conditions distributed over length or area of structural elements as a set of point values of response variables at mesh nodes.

An additional cause of errors – not related to the modeling methodology – is uncertainty in the properties of materials and geometric parameters of structural elements, as well as applied loads, i.e. the uncertainty of the input data. This uncertainty may be associated with stochastic variation of the data, i.e. be aleatory. However, the presence of epistemic uncertainty in the data is also possible.

The resulting set of errors in a finite element model can be represented as a combination:

$$\mathbb{E}_{\mathbf{M}} = \mathbb{E}_{\mathbf{M}a} + \mathbb{E}_{\mathbf{M}c} + \mathbb{E}_{\mathbf{M}d}, \tag{4.27}$$

where $\mathbb{E}_{\mathbf{M}a}$ contains errors in approximating the response variables of the analyzed structure, its properties and applied loads, as well as the boundary conditions within the mesh model, assuming that all quantities included in the calculation are given exact values;

$\mathbb{E}_{\mathbf{M}c}$ covers the errors of computational operations carried out to determine the matrices $[K]$, $[M]$, $[C]$ and the vector column $\{F\}$;

$\mathbb{E}_{\mathbf{M}d}$ is the set of errors in these items of the model due to uncertainty in the input data.

The first two sets are characterized by a predominance of epistemic uncertainty. Possibilities for managing these errors are discussed in Sects. 4.2.4–4.2.6. The third set contains a significant share of aleatory uncertainty, inherited from the uncertainty of experimentally obtained data. Approaches to minimizing this uncertainty are presented in Chap. 5. How the errors belonging to these sets are summed remains generally uncertain and requires special study. However, it is beyond the scope of this book.

In the case of linear static strength analysis, the response of the structure in the form of displacements at the nodes of the finite element ensemble is determined by solving Eq. (4.19):

$$\{U\} = [K]^{-1}\{F\}, \tag{4.28}$$

where $[K]^{-1}$ is the ensemble flexibility matrix (the inverse of the stiffness matrix).

This expression corresponds to the form of the general operator equation (4.1), and we can conclude that:

$$\mathfrak{M} = [K]^{-1}. \tag{4.29}$$

In practice, nodal displacements are rarely determined in this way. Usually, to obtain them Eq. (4.19) is solved by direct methods of linear algebra – such as the Gaussian elimination method or the Cholesky decomposition (most commonly), or iterative methods are used, for example, Seidel method, relaxation method or gradient methods (Zienkiewicz, 1971; Bathe and Wilson, 1976; Faddeev and Faddeeva, 2002). When using any of these methods, a model operator (a computational model in our classification) is determined, and errors $\mathbb{E}_{\mathfrak{M}}$ arise in it.

In the case of using direct methods, which are theoretically exact, these errors can be represented as follows:

$$\mathbb{E}_{\mathfrak{M}} = \Psi(\mathbb{E}_{\mathbf{M}}) + \mathbb{E}_{\mathfrak{M}c}. \tag{4.30}$$

The first term of this combination, $\Psi(\mathbb{E}_{\mathbf{M}})$, denotes the dependence of errors in $\mathfrak{M}$ on the errors in the finite element model, which are described by Eq. (4.27), under the assumption that all computational operations required to obtain the model operator are performed exactly. The second term, $\mathbb{E}_{\mathfrak{M}c}$, accounts for the errors in these operations caused by the representation of numbers in computer memory using a limited number of bits (usually 32 or 64). Because of this, real numbers are stored and processed in computers as imprecise, rounded values, resulting in a loss of precision.

Iterative methods give an approximate solution to Eq. (4.19), which leads to the appearance in the errors of the model operator of a term $\mathbb{E}_{\mathfrak{M}i}$ characterizing them, and in this case the set of errors takes the form:

$$\mathbb{E}_{\mathfrak{M}} = \Psi(\mathbb{E}_{\mathbf{M}}) + \mathbb{E}_{\mathfrak{M}i} + \mathbb{E}_{\mathfrak{M}c}. \tag{4.31}$$

The dynamic response of the structure is determined by numerically integrating Eq. (4.22) over time using step-by-step explicit or implicit methods. In many cases, implicit methods such as the Houbolt, Newmark, or Wilson methods are preferred because of their unconditional stability (Bathe and Wilson, 1976). In the first two methods, to approximate changes in displacements over time, algebraic relations are used that link the sought values of displacements at the moment $t + \Delta t$, $\{U\}_{t+\Delta t}$, with the known values of the kinematic parameters at previous moments in time, which are chosen from the set: $\{U\}_t$, $\{U\}_{t-\Delta t}$, $\{U\}_{t-2\Delta t}$, $\{\dot{U}\}_t$, $\{\ddot{U}\}_t$ (the choice depends on the method used). As a result, ordinary differential equations in the form of Eq. (4.22) are replaced by a sequence of algebraic equations that relate to moments of time increasing with a step within the response analysis interval. The nodal displacements at the end of each new step are determined by solving the equation:

$$[\hat{K}]\{U\}_{t+\Delta t} = \{\hat{F}\}_{t+\Delta t}, \tag{4.32}$$

where $[\hat{K}]$ is the effective stiffness matrix of the finite element ensemble;

$[\hat{F}]_{t+\Delta t}$ is the effective column vector of equivalent nodal forces.

The effective stiffness matrix and the effective column vector of equivalent nodal forces are calculated as follows:

$$[\hat{K}] = [K] + \frac{\alpha_1}{(\Delta t)^2}[M] + \frac{a_2}{\Delta t}[C]; \tag{4.33}$$

$$\begin{aligned}\{\hat{F}\}_{t+\Delta t} = \{F\}_{t+\Delta t} &+ [M]\left\{\varphi_1\left(\Delta t, \left\langle \{U\}_t, \{U\}_{t-\Delta t}, \{U\}_{t-2\Delta t}\right\rangle \text{or} \left\langle \{\dot{U}\}_t, \{\ddot{U}\}_t\right\rangle\right)\right\} \\ &+ [C]\left\{\varphi_2\left(\Delta t, \left\langle \{U\}_t, \{U\}_{t-\Delta t}, \{U\}_{t-2\Delta t}\right\rangle \text{or} \left\langle \{\dot{U}\}_t, \{U\}_t\right\rangle\right)\right\},\end{aligned} \tag{4.34}$$

where α_1, α_2 are coefficients, and φ_1, φ_2 are functions depending on the integration method (Bathe and Wilson, 1976).

In the Wilson method, slightly different expressions are used, but the overall procedure is similar to that described (ibid).

The determination of nodal displacements by solving Eq. (4.32) can be represented in an operator form, similar to static analysis:

$$\{U\}_{t+\Delta t} = [\hat{K}]^{-1}\{\hat{F}\}_{t+\Delta t} = \hat{\mathfrak{M}}\{\hat{F}\}_{t+\Delta t}, \tag{4.35}$$

where $\hat{\mathfrak{M}}$ is the corresponding model operator, $\hat{\mathfrak{M}} = [\hat{K}]^{-1}$.

In practice, the displacements at each time step of the implicit integration procedure are obtained not by inverting the matrix $[\hat{K}]$, but by applying direct or iterative methods of linear algebra to solve Eq. (4.32). In this case, the same errors as those discussed above arise in the model operator that produces the solution to the dynamic problem, but one more term is added to them, which represents the errors in approximating the processes within the framework of numerical integration over time, $\mathbb{E}_{\mathfrak{M}t}$. Thus, the complete set of errors can be written in the following form:

$$\mathbb{E}_{\mathfrak{M}} = \Psi(\mathbb{E}_{\mathbf{M}}) + \mathbb{E}_{\mathfrak{M}t} + \mathbb{E}_{\mathfrak{M}i} + \mathbb{E}_{\mathfrak{M}c} \tag{4.36}$$

(the term $\mathbb{E}_{\mathfrak{M}i}$ comes into play only when using iterative methods).

The errors $\mathbb{E}_{\mathfrak{M}t}$ have been well studied in problems of linear structural dynamics (Bathe and Wilson, 1976). The combination of errors $\mathbb{E}_{\mathfrak{M}t}$ and $\mathbb{E}_{\mathfrak{M}i}$ was the subject of the author's research in connection with one of the problems of nonlinear structural dynamics (Spirochkin, 1987). As above, we will not discuss how the listed errors are summed up and how they are transformed into errors of the response variables. For our purpose, it is sufficient to express the errors in the response by the generalized formula:

$$\mathbb{E}_{\mathbb{R}} = \mathbb{E}_{\mathbb{R}}(\mathbb{E}_{\mathbb{F}}, \mathbb{E}_{\mathfrak{M}}), \tag{4.37}$$

where the term $\mathbb{E}_{\mathfrak{M}}$ is determined according to the Eq. (4.31) or Eq. (4.36).

Below we consider the possibilities of reducing the uncertainty of response variables determined using finite element models by controlling the errors represented by the sets $\mathbb{E}_{\mathbf{M}a}$, $\mathbb{E}_{\mathbf{M}c}$ and $\mathbb{E}_{\mathfrak{M}c}$ that contribute to $\mathbb{E}_{\mathfrak{M}}$.

4.2.4 Refinement of Models by Increasing the Number of Degrees of Freedom

The most common way to reduce approximation errors in mathematical models constructed using FEM is to increase the number of degrees of freedom of the model, n_{dof}. This increase is implemented by the following techniques:

1) increasing the number of response variables considered at each finite element node;
2) increasing the number of nodes in each finite element or some of them;
3) decreasing the element sizes, i.e. increasing the mesh density, which leads to an increase in the number of finite elements in the ensemble.

The first technique means a transition to new types of finite elements that are capable of taking into account additional features of the behavior of the structure under load and the impact of the first cause of errors listed in the previous section. The second technique similarly involves new types of finite elements in the modeling. In these elements, due to additional nodes, higher-order polynomials with a larger number of terms can be used as shape functions, which facilitates to a more accurate approximation of response variables (thereby reducing the errors due to the second cause in the list mentioned above). In addition, it becomes possible to approximate the geometry of curvilinear parts of the structure with a smaller error (see the third cause of errors). The second technique is embodied to the greatest extent in isoparametric finite elements. The third technique, unlike the first two, does not require changing the type of finite elements in the mesh.

The convergence of polynomial approximations used in FEM is theoretically substantiated for elliptic type differential equations. Such equations describe problems in which changes in variables can be reduced to spatial coordinates only, and time can be eliminated from consideration (Tikhonov and Samarskii, 1977). This type of equation is typical for linear static strength analysis. When using shape functions in the form of p-order polynomials to solve a problem described by $2m$-order elliptic differential equations, the relative error in approximating the response variables (such as displacements), can be estimated by the expression (Fried, 1972):

$$\epsilon_a = C\left(\frac{l}{L}\right)^{p+1-m}, \tag{4.38}$$

where C is a constant;
l denotes the characteristic size of the finite element; and
L is the size of the modeled structural element.

Errors of this kind belong to the set $\mathbb{E}_{\mathbf{M}a}$ that appears in Eq. (4.27), enter the set $\mathbb{E}_{\mathfrak{M}}$ and influences the set $\mathbb{E}_{\mathbb{R}}$ determined by Eq. (4.37). In general, the effect of increasing the number of degrees of freedom on approximation errors and, thus, on the refinement of finite element models used to solve various engineering problems described by the above-mentioned differential equations has been well studied (Zienkiewicz, 1971; Strang and Fix, 1973; Zienkiewicz and Morgan, 1983). The management of uncertainty associated with these errors using the techniques listed above is characterized by a fairly high level of development. However, in a number of cases, the idea of increasing the accuracy of mathematical modeling by increasing the number of degrees of freedom loses its effectiveness. This may be the case, for example, when the governing differential equations are not of the elliptic type, due to which the reduction in approximation errors with increasing n_{dof} cannot be estimated using Eq. (4.38). Another typical case is when n_{dof} is increased to such a value that the computational errors when using direct methods of linear algebra (the set $\mathbb{E}_{\mathfrak{M}c}$ in Eq. 4.30), become comparable to or even exceed the expected reduction in approximation errors. In these cases, other methods of refining mathematical models are preferable, such as those described in Sect. 4.2.5, or special control of computational errors is required – see Sect. 4.2.6.

4.2.5 Improving the Approximation of Variables

Polynomial shape functions, most often used in FEM, ensure good convergence for numerical solutions of elliptic type differential equations as the mesh becomes finer. These functions were chosen in the early stages of development of this method, when it was applied mainly to solving linear static strength problems. One of the reasons for this choice was obviously the correspondence of polynomials to exact analytical solutions of differential equations describing bending displacements of extended structural elements, which can be modeled by beams, under loading by concentrated forces. In problems where other loading modes are considered and, along with displacements, other response characteristics are of interest, not to mention the analysis of physical processes of a different nature, polynomials may lose their preference. For example, if we take problems of structural dynamics, they are characterized by distributed inertial loads acting on a structure during vibrations, and vibration frequencies are additional response variables important to designers. In such problems, the mesh density required to approximate all response variables in question with acceptably small errors can lead to very large number of degrees of freedom of the finite element model, and, accordingly, to the number of equations to be solved. At a low rate of convergence of the solution and due to the accumulation of computational errors the technique of refining the model by increasing the number of degrees of freedom, presented in Sect. 4.2.4, turns out to be not so effective. In these cases, alternative techniques become in demand that are capable of reducing the errors in the approximation of the response variables not by increasing n_{dof}, but by using shape functions that are more adequate to the governing differential equations of the problem under consideration. These techniques are demonstrated below on two types of dynamic problems.

The first of them relates to the analysis of low-frequency (sinusoidal) vibrations of a structure, for the modeling of which two-node beam finite elements can be used. Such analyses are conducted, in particular, to evaluate dynamic loads that determine the required strength of the structures of launch vehicles and spacecraft, as well as to form requirements for their control systems at the early stages of design. For this purpose, extended structural elements (fuel lines, stringers, spars, etc. – up to entire booster blocks and spacecraft modules) are modeled by beam finite elements, the longitudinal displacements of which (under tension or compression) are a linear function of the axial coordinate (Eq. 4.3), and the bending displacements are described by third-order polynomials (Eq. 4.6). As calculations carried out for a span of the fuel line of the Russian launch vehicle *Energia* have shown, when modeling this structural element with one beam finite element, the relative errors in determining the natural frequencies of vibrations are (Spirochkin 1987, 1994):

- for the first mode of longitudinal vibrations not less than 10%;
- for the first mode of bending vibrations approximately 20%;
- for the second mode of bending vibrations about 49%.

Enough accurate evaluations of lower vibration frequencies using the beam finite elements of this type (with linear and Hermitian shape functions) can be achieved by modeling a structural element with 5–10 finite elements along its length. In the case of analysis of complex spatial structures containing many elements, for example, the International Space Station, this leads to such an increase in the number of degrees of freedom that it causes large computational costs. They may be acceptable at the final stages of creating a technical system, but they complicate variant analysis at the early stages of design, when many possible configurations are being considered and assessed.

As an alternative, vibrations of structural elements that have the form of simple elastic bodies (a homogeneous beam, a rectangular plate, etc.) could be modeled using special, *modal finite elements* whose shape functions correspond to known analytical solutions that describe the natural vibration modes of such bodies. For example, low-frequency dynamics of a beam-like structural element can be simulated by modeling it by a modal two-node finite element: its longitudinal displacements, u_x, and transverse displacements, u_y, are expressed through the eigenvectors (vibration mode shapes) corresponding to the motion as a rigid body and the first two modes of vibrations as a clamped beam:

$$u_x = \begin{bmatrix} 1 & \cos\frac{\pi x}{l} \end{bmatrix} \begin{Bmatrix} \eta_{x1} \\ \eta_{x2} \end{Bmatrix} = [\Phi_x(x)]\{\eta_x\}; \tag{4.39}$$

$$u_y = \begin{bmatrix} 1 & 1-\frac{2x}{l} & K_4(\lambda_1)K_1\left(\lambda_1\frac{x}{l}\right) - K_3(\lambda_1)K_2\left(\lambda_1\frac{x}{l}\right) & K_4(\lambda_2)K_1\left(\lambda_2\frac{x}{l}\right) - K_3(\lambda_2)K_2\left(\lambda_2\frac{x}{l}\right) \end{bmatrix} \times$$

$$\times \begin{Bmatrix} \eta_{y1} \\ \eta_{y2} \\ \eta_{y3} \\ \eta_{y4} \end{Bmatrix} = [\Phi_y(x)]\{\eta_y\}, \tag{4.40}$$

where η_{x1}, η_{x2}, η_{y1}, ..., η_{y4} are normal (uncoupled) coordinates;

$[\Phi_x(x)]$ and $[\Phi_y(x)]$ are matrices of vibration mode shapes along the x and y axes;

K_1, K_2, K_3, K_4 are Krylov functions, and λ_1, λ_2 are eigenvalues: $\lambda_1 = 4.730$, $\lambda_2 = 7.853$ (Timoshenko and Young, 1974).

The matrices of shape functions (in finite element terminology) for the displacements u_x and u_y in this case are determined as follows:

$$[N_x]=[\Phi_x(x)]\begin{bmatrix}\Phi_x(0)\\ \Phi_x(l)\end{bmatrix}^{-1};[N_y]=[\Phi_y(x)]\begin{bmatrix}\Phi_y(0)\\ \frac{\partial}{\partial x}\Phi_y(0)\\ \Phi_y(l)\\ \frac{\partial}{\partial x}\Phi_y(l)\end{bmatrix}^{-1}. \quad (4.41)$$

Modeling a structural element with just one such modal finite element with the corresponding boundary conditions allows the analysis engineer to obtain values of the frequencies of the first few tones of natural vibrations that are quite close to the exact ones – the error is a fraction of a percent (Spirochkin, 1994). Due to the special type of shape functions, such finite elements do not provide good convergence of the solution when refining the mesh, but this is not required, since they are intended for "one-to-one" modeling, i.e. one structural element should be represented by one finite element.

As a result of using these shape functions to obtain the finite element stiffness and inertia matrices and then incorporating them into the corresponding ensemble matrices, as described in Sect. 4.2.2, a refinement of the finite element model used for dynamic analyses is achieved. The dynamic characteristics of the analyzed structure (natural frequencies and vibration mode shapes) are determined more accurately, as are the kinematic parameters of low-frequency vibrations (displacements, velocities, and accelerations) at points corresponding to the nodes of the finite element mesh. Based on these kinematic parameters, strains and stresses at internal points of each element can be calculated.

It is not always possible, however, to use shape functions that are more adequate to the differential equations describing the problem (and, therefore, more accurate), without introducing additional degrees of freedom into consideration. Since the coefficients in the shape functions are determined through the values of the response variables at the nodes of a finite element, additional nodes may be required to apply the improved functions. In some situations, increasing the number of nodes in a finite element ensemble is difficult because the configuration of the analyzed object itself prevents it. The second type of dynamic problems, which is worth considering, relates to the analysis of coupled vibrations of a structure and a liquid – a problem that arises in the design of a WWER-type nuclear reactor and the development of a system for diagnosing its structural health during operation based on monitoring these vibrations. The problem considered below also gives concrete form to the example of widening the object of analysis presented in Sect. 4.1.3.

As can be seen from Figure 4.2, inside the WWER-1000 reactor vessel, there is the reactor shaft supporting the core. The core is surrounded by the reflection shield that protects the reactor vessel from radiation. The annular gap between the reflection shield and the inner surface of the reactor vessel is very narrow – no more than 35 mm, while sizes of these elements themselves (including the inner diameter of the vessel) are two orders of magnitude larger. This gap is filled with water. To analyze the coupled vibrations of the reactor vessel internals (RVI, including the reactor shaft, reflection shield, core, etc.) and the liquid, it is necessary to take into account the hydrodynamic pressure on the structure and model the volume of liquid in which it occurs. The coupled problem, formulated under the assumption of small structural vibrations, incompressibility, non-viscosity and the absence of a free surface of the liquid, is described by two systems of equations (Zienkiewicz, 1971):

$$[M]\{\ddot{U}\} + [C]\{\dot{U}\} + [K]\{U\} + \{F_p\} = \{F(t)\}; \tag{4.42}$$

$$[H]\{p\} + [Z]\{\ddot{U}\} = 0, \tag{4.43}$$

where $\{F_p\}$ is the column vector of nodal forces equivalent to the hydrodynamic pressure:

$$\{F_p\} = \frac{1}{\rho^*}[Z]^{\mathrm{T}}\{p\}; \tag{4.44}$$

$[H]$ is a matrix characterizing the reaction of a liquid to an external mechanical impact in the form of values of the hydrodynamic pressure $\{p\}$ at the nodes of a finite element mesh covering the volume of liquid, V:

$$[H] = \sum \int_V [N]^{\mathrm{T}}[N]dV; \tag{4.45}$$

$[Z]$ is a matrix that relates the values of the hydrodynamic pressure at the nodes on the wetted surface of the structure (the interface area between the structure and the liquid, Ω), with the accelerations at the nodes of the finite element model of the structure:

$$[Z] = \sum \int_\Omega [N]^{\mathrm{T}}\rho^*[N_b]d\Omega; \tag{4.46}$$

ρ^* is the density of the liquid.

Here $[N]$ is the matrix of the shape functions for the pressure distribution over the volume of a finite element modeling the liquid and the matrix $[N_b]$ relates nodal accelerations of the element facet belonging to the interface area between the structure and the liquid with the acceleration in the direction of the normal. The matrices $[Z]$ and $[H]$ are formed in the usual way from the corresponding finite element matrices $[H]_e$ and $[Z]_e$.

The last three relations allow us to reduce the coupled problem to the following equation:

$$([M] + [M^*])\{\ddot{U}\} + [C]\{\dot{U}\} + [K]\{U\} = \{F(t)\}, \tag{4.47}$$

where $[M^*]$ is the inertia matrix for the added mass of the liquid,

$$[M^*] = -\frac{1}{\rho^*}[Z]^{\mathrm{T}}[H]^{-1}[Z]. \tag{4.48}$$

For adequate approximation of hydrodynamic pressure gradients in the liquid-filled annular gap, it should be divided into several layers of simple three-dimensional finite elements (tetrahedrons or hexahedrons) along the radial coordinate, or isoparametric elements with internal nodes should be used. The sizes of the elements along the other two coordinates (circumferential and vertical) should not differ significantly to avoid large numerical errors when determining the above matrices. As a result, the required discretization turns out to be excessive for an adequate description of coupled vibrations in the low-frequency range, which is of primary interest for designers, and the computational costs become very large. The increase in computational costs is amplified by the fact that the matrix $[M^*]$, unlike the matrix of inertia of the structure, $[M]$, is not sparse. To solve this issue, special finite elements were developed that are capable of modeling thin layers of liquid filling narrow gaps (Spirochkin, 1997, 1998, 2011). These elements combine the minimum degree of required discretization (it is possible to use one layer of elements along the gap width) with high accuracy of approximation of the change in hydrodynamic pressure. For their construction, two procedures were proposed that allowed improving the pressure shape functions.

The first procedure involves introducing additional internal nodes into the finite element in the form of a hexahedron with eight nodes at its corners (H8) and determining a matrix $[H]'_e$ for the constructed multi-node element, which can be represented as:

$$[H]'_e = \begin{bmatrix} H_{cc} & H_{ca} \\ H_{ca}^{\mathrm{T}} & H_{aa} \end{bmatrix}, \tag{4.49}$$

where H_{cc} and H_{aa} are submatrices corresponding to the degrees of freedom (pressure values) at corner nodes and additional nodes, respectively, and H_{ca} is a cross matrix characterizing the relationship between the first and second degrees of freedom. In fact, the matrix $[H]'_e$ itself is not determined, but these submatrices are calculated. They are then involved in the reduction of this matrix to the original (corner) nodes according to the well-known algorithm (Zienkiewicz, 1971), the result of which is the refined matrix $[H]_e$:

$$[H]_e = H_{cc} - H_{ca}H_{aa}^{-1}H_{ca}^{\mathrm{T}}. \tag{4.50}$$

The improvement of pressure shape functions using such a reduction can be expressed by the formula:

$$[N] = N_c - N_a H_{aa}^{-1} H_{ca}^{\mathrm{T}}, \tag{4.51}$$

where N_c and N_a are submatrices related to the corner and additional degrees of freedom of the finite element, respectively.

Additional nodes can be introduced either by dividing the original finite element into several layers of smaller elements, as shown in Figure 4.4, or by inscribing a multi-node isoparametric element into its boundaries. The element obtained in this way is designated as H8R.

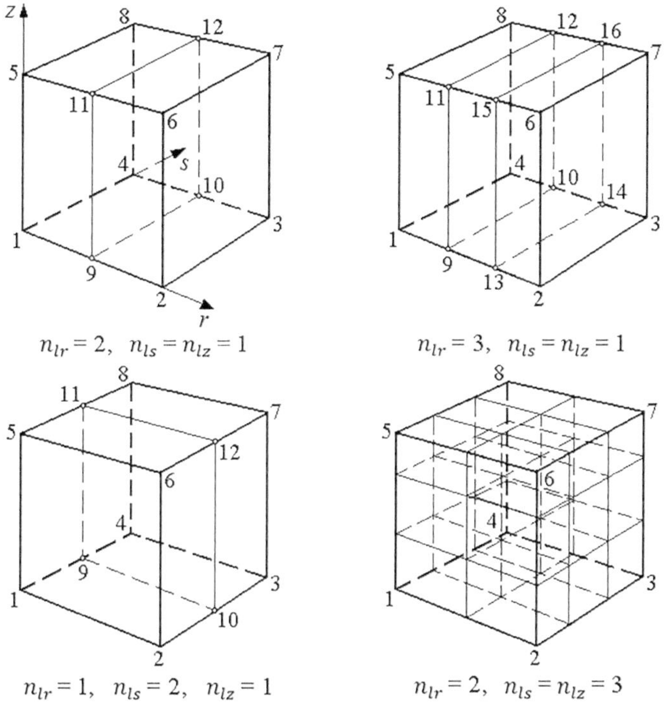

Fig. 4.4 Construction of finite elements of the liquid by introducing additional internal nodes (the axis r coincides with the radius of the reactor vessel, the variables n_{lr}, n_{ls}, n_{lz} denote the number of layers into which the original element is divided).

This procedure results in an approximation of the pressure distribution inside the H8R finite element by a sequence of piecewise linear shape functions, typically applied to hexahedrons:

$$N_i = \frac{1}{8}(1 + \tilde{r})(1 + \tilde{s})(1 + \tilde{z}), \tag{4.52}$$

where i is the node number in such an element ($i = 1, 2, \ldots 8$); $\tilde{r} = rr_i$, $\tilde{s} = ss_i$, $\tilde{z} = zz_i$ ($r_i = \pm 1$, $s_i = \pm 1$, $z_i = \pm 1$).

In the original H8 element, such functions are defined inside its volume, while in the element divided into small internal hexahedrons – inside each hexahedron. In the case of inscribing a multi-node isoparametric finite element into the H8 element, the pressure approximation is performed by higher-order shape functions (Spirochkin, 1998).

The second procedure for improving the pressure shape functions is to use analytical solutions of the governing differential equations describing the problem at hand. For example, it is possible to construct special finite elements for the gap between a cylindrical shell and a rigid wall based on the solution of the Laplace equation in cylindrical coordinates r, φ (which is valid under the assumption of an infinite length of the shell and liquid along the axial coordinate):

$$\frac{1}{r}\frac{\partial}{\partial r}\left(r\frac{\partial p}{\partial r}\right)+\frac{1}{r^2}\frac{\partial^2 p}{\partial \varphi^2}=0 \tag{4.53}$$

under the boundary conditions:

$$\frac{\partial p}{\partial r}=-\rho^*\frac{\partial^2 w}{\partial t^2}\ \text{at}\ r=R_1;\ \frac{\partial p}{\partial r}=0\ \text{at}\ r=R_2, \tag{4.54}$$

where R_1 is the radius of the shell and R_2 is the inner radius of the rigid wall; w is the radial displacement.

The solution of this equation, written as an interpolation formula for determining the pressure in the gap between the shell and the rigid wall, has the following form:

$$p=\varphi(r)p_1+[1-\varphi(r)]p_2, \tag{4.55}$$

where p_1 and p_2 are the pressure values on the shell surface and the wall respectively; $\varphi(r)$ is a function determined by the expression (Spirochkin, 1998):

$$\varphi(r)=\frac{\left(\frac{r}{R_2}\right)^n+\left(\frac{r}{R_2}\right)^{-n}-2}{\left(\frac{R_1}{R_2}\right)^n+\left(\frac{R_1}{R_2}\right)^{-n}-2}. \tag{4.56}$$

In this expression, n is the number of waves of bending vibrations of the shell along its circumference, and r varies in the range from R_1 to R_2.

This procedure is implemented in a special isoparametric finite element of liquid (H8B) with a rigid boundary facet corresponding to the contact of the liquid with a rigid wall. When analyzing coupled vibrations of RVI, the narrow annular gap between the reflection shield and the inner surface of the reactor vessel filled with water, as well as other thin layers of liquid, can be modeled within the thickness by a single layer of H8R or H8B finite elements. In other directions, the volume of liquid can be meshed based only on the required degree of discretization of the part of the structure that interacts with it.

Both procedures were verified using the example of coupled vibrations of a cylindrical shell located inside a rigid cylindrical channel with liquid (see Figure 4.5), for which an approximate analytical solution based on the Flügge shell theory is known (Chen and Rosenberg, 1975).

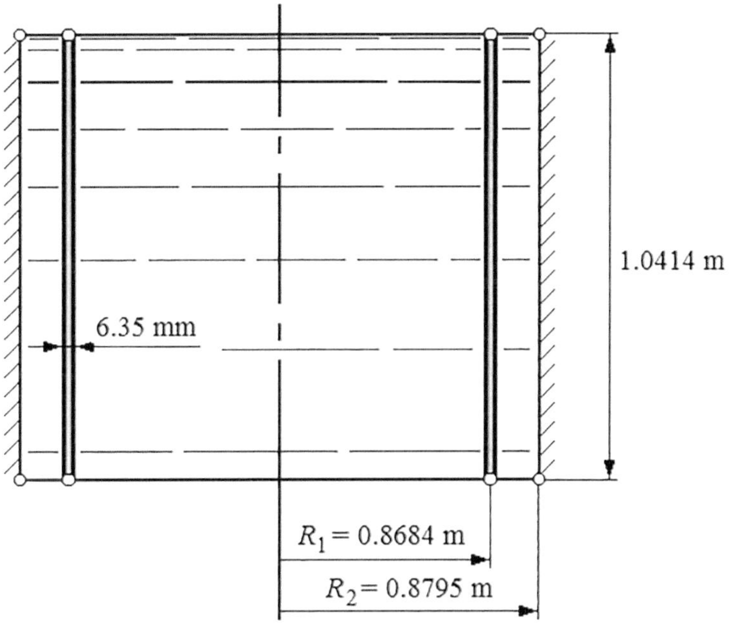

Fig. 4.5 Verification example: cylindrical shell in a rigid cylindrical channel filled with liquid.

The properties of the shell material are as follows:

Young's modulus $E = 1.87 \cdot 10^{11} \dfrac{\text{N}}{\text{m}^2}$;

Poisson's ratio $\mu = 0.27$;

density $\rho = 7990 \dfrac{\text{kg}}{\text{m}^3}$.

Inside and outside the shell there is a liquid: $\rho^* = 922.6 \dfrac{\text{kg}}{\text{m}^3}$.

The boundary conditions for the liquid at the ends of the cylindrical volume correspond to zero pressure.

Verification was carried out by calculating the natural frequencies and mode shapes of the coupled vibrations of the shell, i.e. solving the eigenvalue problem corresponding to Eq. (4.47), in which damping was neglected, and comparing them with the known solution. A finite element model of one half of the shell and adjacent volumes of liquid was used for the calculation. The shell was modeled by four-node plate elements (P4). Several mesh variants for the half of the shell were considered: they corresponded to dividing it into 16, 32 and 64 elements along the circumference and into 5 and 8 layers of elements in the vertical direction. One of these variants, including a shell model containing 16×5 elements, as well as corresponding models of the volumes of liquid inside and outside it, is shown in Figure 4.6. To model the volume of liquid inside the shell, it was divided in the radial direction into 2 layers of H8 or H8R elements and 2 layers of prismatic (wedge-shaped) finite elements with 6 nodes in the corners (P6). The liquid in the narrow gap between the shell and the rigid wall of the channel was modeled by a single layer of H8, H8R or H8B elements along the radius.

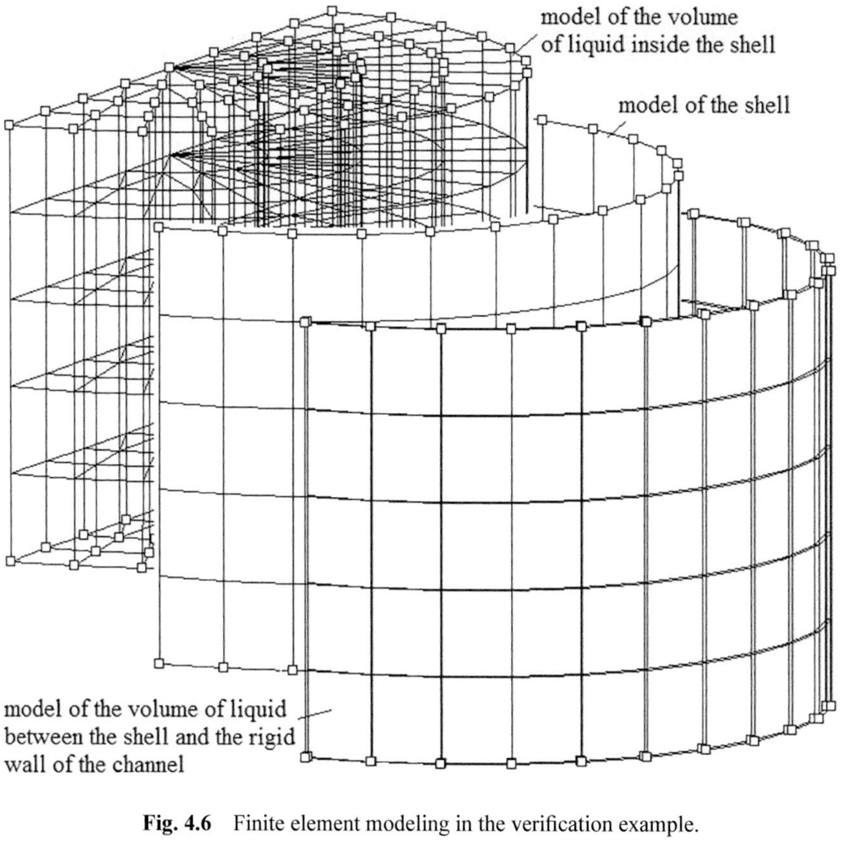

Fig. 4.6 Finite element modeling in the verification example.

As a result of the numerical solution of the eigenvalue problem, dependencies were obtained between the frequency of coupled vibrations, *f*, and the number of bending waves around the circumference of the shell, *n*, for different mesh variants. These dependencies are depicted in Figure 4.7 in comparison with reference data (Chen and Rosenberg, 1975).

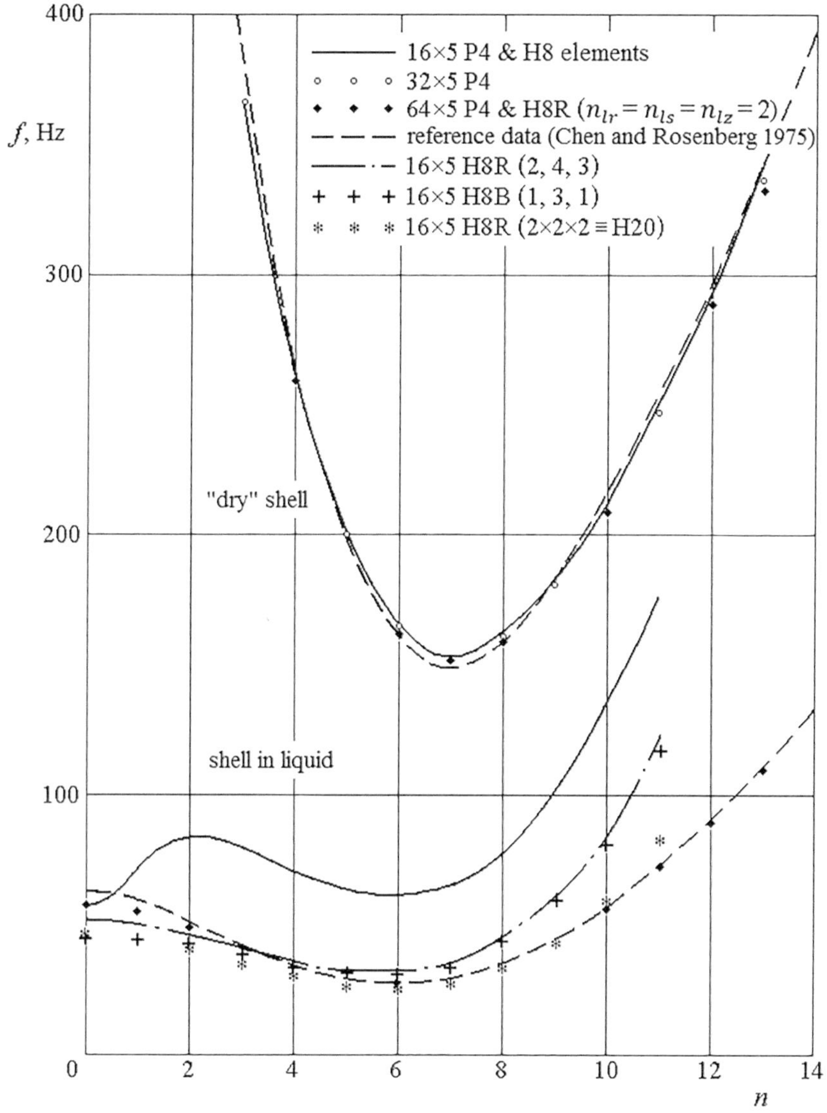

Fig. 4.7 Frequency of coupled vibrations, *f*, vs. the number of bending waves around the circumference of the shell, *n*, for different mesh variants in the verification example.

As the curves corresponding to the "dry" shell show, increasing the degree of its division around the circumference from 16 to 64 finite elements leads to a slight refinement of the natural frequencies. Their values remain almost unchanged when the number of finite element layers in the vertical direction increases from 5 to 8. Thus, the 16×5 mesh variant applied to one half of the shell is quite acceptable in terms of accuracy.

When analyzing shell vibrations in the liquid and modeling the latter with simple H8 elements, the calculated natural frequencies differ significantly from reference values. The largest error is observed in the frequency of the lowest mode associated with 6 bending waves along the circumference, and for the 16×5 shell mesh variant, this error is more than 100%. Such a high error is explained by both the insufficient degree of discretization of the liquid in the gap between the shell and the wall, and the ratio of the sizes of the elements modeling it (1:15:19), which is unacceptable when using shape functions according to Eq. (4.52).

Replacing simple H8 hexahedrons with special H8R or H8B finite elements with improved shape functions allows the analysis engineer to noticeably improve the accuracy of determining natural frequencies (see Figure 4.7). Depending on the choice of one or another type of elements and the degree of internal dividing (in the case of reduction), it is possible to obtain estimates of the lower frequencies both "from above" (using a multilayer ensemble of small H8 elements in H8R) and "from below" (modeling by H8B elements or inscribing H20 elements into H8R).

The main parameters of the H8R finite elements, which can be used to control the accuracy of approximation of hydrodynamic pressure in a narrow gap and, accordingly, the accuracy of determination of the shell vibration frequency in the liquid, are the indicators of internal dividing the initial H8 elements along the r and s axes: n_{lr} and n_{ls}. The minimum error in the value of the frequency of the lowest mode of coupled shell vibrations ($n = 6$) in the studied range of n_{lr} and n_{ls} is achieved for the H8R elements at $n_{lr} = 2$ and $n_{ls} = 4$ is about 12.9% (internal dividing into 3 layers is performed along the z axis). When modeling the gap using the H8B finite elements ($n_{lr} = 1, n_{ls} = 3, n_{lz} = 1$), the error in the value of the lowest mode frequency decreases to 7.3%, but the errors in the values of frequencies corresponding to the interval $0 \leq n < 3$ increase. In all cases, the modeling of the layers of liquid adjacent to the shell inside it was performed using H8R elements with the indicators of internal dividing: $n_{lr} = n_{ls} = n_{lz} = 3$.

The considered verification example demonstrates the applicability of the described procedures of improvement of the shape functions for the analysis of coupled vibrations of structures interacting with thin layers of liquid.

Both techniques presented above, which involve the construction and use of special finite elements (modal for a structure or hydrodynamic for a liquid) with improved shape functions, contribute to reducing the approximation errors corresponding to $\mathbb{E}_{\mathbf{M}a}$ in Eq. (4.27). By restraining the uncontrolled increase in the number of degrees of freedom of the finite element ensemble, these techniques also limit the errors of the computational operations, $\mathbb{E}_{\mathbf{M}c}$. They were implemented in the *NewTone* software system, developed by the author and his colleagues to analyze structural dynamics problems (Spirochkin, 1993, 1994, 1997, 1998).

Along with these techniques, there are other approaches to improving the description of response variables. For example, the capabilities of FEM in solving particularly complex dynamic problems can be expanded by using:

- macroelements (multi-node elements without an established internal structure), in which the variables are described on the basis of experimental data (Spirochkin, 1987, 1993; Du Bois et al., 2000; Abramowicz, 2001);
- spectral elements that use approximation of variables by Legendre polynomials (Levin and Vershinin, 2015);
- meshfree approximations (Belytschko et al., 1996; Huerta et al., 2004; Tang et al., 2016; Chen et al., 2017).

4.2.6 Controlling Computational Errors

Errors in computational operations, which are characterized by the set $\mathbb{E}_{\mathfrak{M}c}$ and affect the errors of the model operator, $\mathbb{E}_{\mathfrak{M}}$, according to Eq. (4.30), can bring significant uncertainty into the values of response variables determined by calculations using finite element models. This uncertainty is expressed through the set of response errors, $\mathbb{E}_{\mathbb{R}}$, – see Eq. (4.37). Computational errors arise from the limited number of bits in the representation of numbers in computer memory and, as a consequence, the rounding of the results of calculations with real numbers. These errors manifest themselves mainly when performing arithmetic operations with numbers whose order differs greatly from each other. Such operations are typical for direct methods of solving linear algebraic equations obtained within the framework of FEM, including the application of the Gaussian elimination method and the Cholesky decomposition for Eq. (4.19) or Eq. (4.32) (Bathe and Wilson, 1976).

The level of computational errors due to rounding of real numbers, ϵ_c, is determined by the relative error of representation of these numbers in a computer system, ϵ_n, and the degree of conditionality of the stiffness matrix $[K]$ (or the effective stiffness matrix $[\hat{K}]$), expressed by the condition number. The conditionality of the matrix $[\hat{K}]$, determined according to Eq. (4.33), does not deteriorate compared to the conditionality of the stiffness matrix appearing in this equation, therefore, in order to estimate the computational errors, we can consider the condition number of the latter, *cond* $[K]$, and as the desired estimate we obtain:

$$\epsilon_c \geq \epsilon_n \, cond\,[K]. \tag{4.57}$$

The relative error in representing real numbers in a computer system, ϵ_n, depends on the number of digits allocated for storing the mantissa, n_{dsm}, – the larger this number, the smaller the error:

$$\epsilon_n = 2^{-(n_{dsm}+1)}. \tag{4.58}$$

The condition number of the matrix [*K*] can be estimated in various ways, e.g., as the ratio of the largest and smallest of its eigenvalues λ_i (Faddeev and Faddeeva, 2002):

$$cond[K] = \frac{max|\lambda_i|}{min|\lambda_i|}, i = 1, 2, \ldots n_{dof}. \tag{4.59}$$

The condition numbers of the stiffness matrices of the models used when substantiating the strength of the structures whose elements differ significantly in their stiffness characteristics can be of the order of tens of thousands or more. This gives an idea of how large the computational errors belonging to the set $\mathbb{E}_{\mathfrak{M}c}$ can be. These errors have a corresponding effect on the set $\mathbb{E}_{\mathbb{R}}$ determined by Eq. (4.37), increasing the errors of the obtained response variables. In the case of using direct methods, they can be quite significant. Due to this, along with large expenditures of time and computer memory, these methods are used when the number of algebraic equations to be solved does not exceed several hundred thousand (Levin and Vershinin, 2015). When using iterative methods (the most advanced of which do not require storing the ensemble stiffness matrix) a large condition number leads to slow convergence of the solution, i.e. also has a negative effect.

Thus, special control of computational errors related to the condition number of the ensemble stiffness matrix (or other matrices that play a similar role in the model operator) is required. It can include the use of:

- computers whose architecture ensures the highest (among available computing systems) precision in representing real numbers, i.e. the lowest level of errors ϵ_n;
- methods of coding variables in the FEM application program that implement this precision;
- algorithms that reduce the accumulation of rounding and truncation errors in calculation results;
- special methods of reducing the condition number.

Among the latter, it is worth mentioning an engineering technique such as aligning the finite element mesh so that the sizes of the finite elements in different directions do not differ greatly from each other. Furthermore, one can resort to smooth mesh refinement in areas where the mechanical properties and geometric parameters of the analyzed structure change significantly. It is also possible to express the degrees of freedom corresponding to members of the stiffness matrix that stand out sharply from the average level through other degrees of freedom, provided a kinematic relationship can be established between them, etc.

Mathematical methods for controlling computational errors that arise due to a high condition number include preconditioning of the stiffness matrix. Preconditioning is mainly used to solve Eq. (4.19) or Eq. (4.32) by iterative methods (Levin and Vershinin, 2015).

4.3 Analytical Reconstruction Based on Available Fragments of Information

4.3.1 Definition of Analytical Reconstruction and Its Application in Design

In the context of the issues that we consider, analytical reconstruction is the restoration of lost information characterizing an object of interest (a technical system, an applied load, a response, etc.), or the acquisition of hidden information about it using mathematical methods for processing available fragments of this information. The goal of such reconstruction is to reduce epistemic uncertainty caused by incomplete knowledge of the object by obtaining previously unknown data about its properties, structure, state, or behavior.

Analytical reconstruction is in demand when solving a number of design problems. One such problem is determining the field of a variable based on its experimentally recorded values. Such experiments are conducted, for example, on mock-ups, small-scale physical models, or prototypes of the designed system to determine aerodynamic, acoustic, or thermal loads. Vibration tests can also be mentioned, which are carried out to confirm the system's strength and operability, as well as to determine its dynamic characteristics. The latter are used to update the mathematical model of the designed system used during its operation, and to fine-tune the control system and vibration diagnostic system. In experiments, unlike mathematical simulations, it is difficult to obtain data on the variable in question at any point of the system being tested, since it is impossible to install sensors everywhere, or the required number of sensors becomes so large that their installation and processing of the obtained data entails unacceptable costs. As a result, one has to limit oneself to recording data at a small number of points and, based on this data, construct the field of the variable, i.e. a function describing the distribution of its values, for the entire system. Thus, this problem falls under the category of secondary processing of experimental data with the aim of interpreting them as a whole and making them applicable to any element of the system.

Another problem that may require analytical reconstruction is determining the loads acting on a system in certain operating modes based on recorded data characterizing the response to these loads. This problem arises when direct load recording is impossible or difficult, for example, when an aircraft is exposed to atmospheric turbulence or impact forces from the landing surface. In such cases, kinematic response parameters, typically velocity or acceleration, are recorded at one or more points on the aircraft, and loads are calculated from them (see Sect. 2.1.2). Typically, such studies are conducted using prototypes or experimental specimens of a system, and the results are transferred to new systems or subsequent production series. Load reconstruction may also be required when analyzing extreme events leading to complete or partial system collapse. The input data for this analysis are observed irreversible deformations and other evidence of damage. The purpose of the analysis is to improve existing systems and update maximum load values for use in future projects.

The third problem associated with the concept of reconstruction is reengineering, i.e. the creation of a system similar to some prototype, about the construction of which there is no complete set of data, but only fragmentary information, often limited to a description of its functioning. Reengineering is manifested in *bionics*, i.e., in the development of technical systems based on principles observed in living nature, in the reproduction of advanced technological achievements of a competitor or adversary, as well as in the recreation of lost artifacts of technical progress.

The corresponding directions of reconstruction are discussed in more detail below. Analytical reconstruction is also used in common engineering tasks, such as determining the internal state of a part of a structure based on known values of response variables at its boundaries, or the dynamic load profile over time based on its stationary spectrum or shock spectrum, and so on.

4.3.2 Determining the Field of a Variable Based on Its Experimentally Obtained Values at a Number of Points

Reconstruction of the a priori unknown field of a variable, Y (the distribution of its values over spatial coordinates and (or) in time), about which there are only fragmentary experimental data, Y_1, Y_2, ... Y_{N_Y} (N_Y is the number of known values), can be performed by approximating it using a set of functions, ψ_1, ψ_2, ... ψ_{N_ψ} well fitted to the specific problem:

$$Y = w_1\psi_1 + w_2\psi_2 + \ldots + w_{N_\psi}\psi_{N_\psi}, \tag{4.60}$$

where w_1, w_2, ... w_{N_ψ} are weighting coefficients;

N_ψ is the number of functions used as *constructs*.

The expression "well fitted" means that the approximating functions correspond to the physical process under consideration and the differential equations describing the problem, and also yield values of the variable that are exactly equal to the values Y_1, Y_2, ... Y_{N_Y} at the spatial and temporal points where these experimental data are recorded. If in time the recordings are usually made at a certain interval, then in space the recording points can be located arbitrarily, without forming any regular layout.

For example, to approximate the spatial distribution of vibration displacement amplitudes obtained during vibration test of a system, polynomials or trigonometric functions determined for its entire structure are applicable: $\psi_i = \psi_i(x, y, z)$. The weighting coefficients can be determined exactly in the case when $N_\psi = N_Y$:

$$\begin{Bmatrix} w_1 \\ w_1 \\ \cdots \\ w_{N_\psi} \end{Bmatrix} = [Q]^{-1} \begin{Bmatrix} Y_1 \\ Y_2 \\ \cdots \\ Y_{N_Y} \end{Bmatrix}, \tag{4.61}$$

where $[Q] = [\psi'_{ij}]$ is a matrix of values of various constructs at points where the values of Y are known; $i = 1, 2, \ldots N_Y$; $j = 1, 2, \ldots N_\psi$.

If N_ψ is not equal to N_Y, then the weighting coefficients are determined approximately, for example, using the least squares method.

When a system with a complex spatial configuration is subjected to experimental study, as well as in the case of significant differences in the geometry and properties of its elements, it is preferable to use constructs ψ_1, ψ_2, ... ψ_{N_ψ}, the domains of definition of which are individual parts, and not the entire structure as a whole. Such approximation technique, based on locally defined functions, is similar to FEM, when the field of the variable is the desired result of modeling. This technique is implemented on a finite element mesh, the main nodes of which coincide with the points where experimental data, Y_1, Y_2, ... Y_{N_Y}, are recorded, and the nodal variables are the values of these data. When using one or another approximation technique, the value of the variable in question at any point of the structure can be found by interpolation according to Eq. (4.60) with the calculated weighting coefficients.

Ideally, the sensor installation layout should cover the entire domain of the subsequent reconstruction, ensuring reliable interpolation. This layout should be as regular and uniform as possible, especially when using finite element approximations, to allow for the generation of an acceptable mesh of triangles and quadrilaterals, for which well-developed sets of approximating functions (shape functions) exist.

The task becomes more complicated when any of these requirements is not met – then additional processing of experimental data is necessary. If the reconstruction domain extends beyond the sensor installation area, then the field of the variable Y can be determined by extrapolation. If the sensor installation layout is significantly irregular or non-uniform, then intermediate interpolation can be used to fill large spatial gaps, providing appropriate "multiplication" of the experimental data. Both operations, interpolation and extrapolation, bring errors into the reconstruction result; this process generally correspond to source 6 in Table 3.1.

Thus, the described direction of reconstruction is connected with the methods of the *design of experiments* (Fisher, 1971; Montgomery, 2013; Antony, 2014; Durakovic, 2017), and this connection should be taken into account when solving the corresponding problems.

4.3.3 Determining the Loads Acting on a System Based on Its Response

As noted in Sect. 4.3.1, determining the loads acting on a system based on experimental data on the system's response to these loads may be necessary both for normal operating conditions and for extreme events that lead to system collapse. Loads are determined through calculations using mathematical models that describe loading processes and the behavior of the structure under loads. The models are constructed primarily using FEM.

In normal operating conditions characterized by static or quasi-static loading (absence of structural vibrations), linear models corresponding to Eq. (4.19) are applicable. If the loads are dynamic, i.e. they change over time, causing elastic vibrations of the structure, then Eq. (4.22) is used. Since the amount of experimental data on the system's response is typically limited, the problem of determining the loads causing a given response generally does not have a single solution. To minimize this uncertainty and obtain a single, adequate solution, additional data on the nature of the load distribution across the structure's surface and its time profile are required. With such data, the task of load reconstruction can be reduced to determining numerical values that act as scaling factors in the corresponding loading function.

The ambiguity of load reconstruction increases for extreme events that lead to structural collapse. In the case of loads that change rapidly over time, especially impact loads, the choice of a mathematical model that adequately describes high-speed deformation processes is of crucial importance. At certain strain rates and due to the influence of elastic-plastic wave propagation within the structure on the dynamic response, Eq. (4.22) becomes inapplicable and other models should be used (Zukas et al., 1982). Any of the potentially more accurate models must be validated, and the results of computer simulations verified (Oberkampf et al., 2003; Oberkampf and Roy, 2010; Roy and Oberkampf, 2011).

Thus, the reconstruction of loads, on the one hand, reduces the epistemic uncertainty corresponding to source 6 (see Table 3.1), and, on the other hand, is capable of increasing it.

4.3.4 Reengineering

Reengineering in the form of developing technical systems that replicate the principles of living nature, has been a part of design practice for a very long time, long before the emergence of the engineering discipline known as "bionics". For example, Leonardo da Vinci invented a flying machine that implemented the ornithological principles of flight characteristic of birds.

Bionics in its current state is a vast interdisciplinary field of R&D. For an introduction to this field, we refer the reader to the relevant literature. Here, we note that the creation of AI in the form of *neural networks* that reproduce, according to modern views, the structure and operating principles of the human brain, can be considered one of the areas of bionics and reengineering. We will discuss some issues related to bionic design and relevant to the topic of this book in Sects. 6.5.3, 7.5, 8.1 and 9.3.1.

Reengineering is often used in high-tech areas, including defense and deterrence, to reproduce the advanced technologies of a competitor, real or potential adversary. For example, modern space technology owes much of its early development to the Soviet and US reconstruction of the original German achievements implemented in the V-2 rocket in the 1940s, which the Allies obtained after capturing the Peenemünde research center and the rocket production facilities. The trophies proved to be quite useful and timely, as the author can judge from the Soviet experience, since in the USSR there were already groups of engineers and scientists conducting similar independent work in this area. These specialists possessed the necessary qualifications to understand the German experience, despite the fragmentary nature of the materials obtained, and to further develop rocket technology.

During the era of superpower rivalry, Soviet and American specialists working in rocketry and astronautics, aviation, nuclear engineering, computing, and related fields resorted to reconstructing the secret developments of their foreign competitors to keep up in the technological race. The data used to reconstruct cutting-edge developments was often the result of technical espionage. Besides being fragmentary, it sometimes contained deliberate disinformation intended to mislead a competitor. It is unlikely that the situation has fundamentally changed today; the ongoing changes may only affect a circle of competitors and a specific high-tech industry, the latest developments in which each participant views as a matter of national prestige and (or) national security.

Recreating lost artifacts of technical progress sometimes arises as a design task, involving the restoration of missing design documentation. Such tasks represent a less security-sensitive area of reengineering, as they mainly concern history, including the history of technology, archaeology, and some other non-military fields of knowledge. One example is the study and reconstruction of the Antikythera Mechanism to understand its structure, part of which was missing, its purpose, and its operation (Mavromanolakis et al., 2019).

For successful reengineering, both the availability of the necessary, albeit fragmentary, data and clarity regarding the purpose, configuration and functioning of the reconstructed object are important. Therefore, the essence of this direction of reconstruction can be expressed as a generalized mathematical relationship similar to Eq. (4.60), in which the terms are certain mathematical entities corresponding to the reconstruction task and the nature of the object, as well as the available fragments of information, Y_1, Y_2, ... Y_{N_Y}. In particular, if the reconstruction consists of restoring the geometry of an object, these entities are suitable geometric bodies. Questions about the minimum number of fragments required for reconstruction, their sizes, and how these fragments should relate to each other remain generally open. For example, the problem of reconstructing a giant statue at the Temple of Hercules in Amman, Jordan (presumably a statue of Hercules), based on two surviving fragments—a hand and a knee (Figure 4.8)—seems at first glance insoluble.

However, despite the obvious extreme uncertainty inherent in this problem, it is fundamentally possible to resolve it with the acquisition of additional information. Such information could include, for example, geometric relationships between parts of the human body and its dimensions as a whole, which were used by ancient sculptors, descriptions or images of this statue or similar ones in historical sources, canons of sculpting gods and heroes in the era of the statue's creation, and so on.

Following these considerations, we come to the need to develop general principles and applied methods of analytical reconstruction aimed at reducing the epistemic uncertainty associated with the unknown characteristics of objects with which designers have to deal. For such a development, it may be helpful to turn

Fig. 4.8 Fragments of a statue at the Temple of Hercules in Amman. Photo by K. Mango. https://commons.wikimedia.org/wiki/Category:Amman_Citadel#/media/File:Hand_of_Hercules.jpg, CC BY-SA 4.0.

to group theory and the McKay conjecture. According to this conjecture, which pertains to finite groups[1] (McKay, 1972), many important properties of a group can be determined by examining only a fragment of the group constructed in a particular way. The possibility of such a determination is based on a fundamental correspondence between the group and the fragment. When applied to functions, it is often sufficient to limit oneself to analyzing the function's behavior for a small set of input values to understand its behavior for any possible parameters. McKay substantiated his conjecture for certain simple groups. A proof of the conjecture for the general case appeared in late 2024 (Cabanes and Späth, 2024).

The conjecture in question can be interpreted as follows. If an acceptable mathematical description of a fragment of a physical or mathematical entity, such as a field of a variable, is found, and the assumption is valid that the factors shaping or influencing this field continue to act beyond the fragment under study, then interpolation or extrapolation based on the found description can be used to obtain the values of the variable at any spatiotemporal point.

4.4 Other Applications of Analytical Methods and General Assessment of the Applicability of the Analytical Approach to Uncertainty Reduction

4.4.1 Other Applications

Among other applications of analytical methods to reduce uncertainty in the design of technical systems the following should be mentioned:

[1] A finite group contains a finite number of elements.

- elimination of errors and noise from experimental data and calculation results;
- substantiation of calculation results in the absence of experimental data;
- modeling of human factors.

In our understanding, errors in experimental or calculated data are point deviations or spatial distributions of deviations from a priori unknown correct values of the response of the system under study. Noise is a process that pollutes the useful signal, i.e., the function of change of the response variable of interest over time. These errors and noise are considered below as manifestations of the sixth source of uncertainty in the classification presented in Table 3.1.

Errors in experimental data can relate to both aleatory and epistemic uncertainty components. To manage these components, appropriate methods of the design of experiments, analysis and processing of experimental data, such as those described in the publications cited in Sect. 4.3.2, are applicable. Errors in calculations performed using finite element models were discussed in Sects. 4.2.3, 4.2.4 and 4.2.6. An example of such errors is discontinuities (jumps) in the values of internal forces and stresses at the boundaries of adjacent finite elements, which arise due to the approximation of displacements within each element using polynomials of insufficiently high order. To eliminate these jumps, procedures of artificial "smoothing" of the calculation results are used.

A more correct technique from a physical point of view is to refine the calculated response values by taking into account in each finite element additional displacements caused by the distributed load and not accurately described by conventional shape functions. In particular, for two-node beam finite elements, whose axial and lateral displacements are described by Eqs. (4.5) and (4.6), additional displacements $\tilde{u}_x$ and $\tilde{u}_y$ are introduced, caused by distributed inertial forces arising from vibrations. They correspond to particular solutions of the differential equations for tension-compression and beam bending:

$$\frac{\partial^2 \tilde{u}_x}{\partial x^2} = \frac{\rho}{E} \ddot{u}_x; \tag{4.62}$$

$$\frac{\partial^4 \tilde{u}_y}{\partial x^4} = -\frac{\rho A_{cs}}{EI_{cs}} \ddot{u}_y, \tag{4.63}$$

where A_{cs} and I_{cs} are the area and moment of inertia of the beam cross section, respectively.

Particular solutions are determined for the following boundary conditions:

$$\tilde{u}_x = \tilde{u}_y = \frac{\partial \tilde{u}_y}{\partial x} = 0 \text{ at } x = 0, x = l, \tag{4.64}$$

i.e., additional displacements do not affect the general solution in the form of displacements at the nodes of the finite element, satisfying the equilibrium equations.

Expressions for the additional displacement shape functions are obtained by integrating Eqs. (4.62) and (4.63) with these boundary conditions. For axial displacements, the order of the leading term of the approximating polynomial is three, and for lateral displacements, it is 7 (Spirochkin, 1994). As a result of introducing additional displacements, jumps in the values of internal forces upon transition across the boundary between two adjacent elements disappear. This technique can be applied to any other types of finite elements.

Noise present in the results of experiments and calculations can have various origins. In experiments, it is caused primarily by the stochastic variation of physical factors and is characterized by aleatory uncertainty. In computer simulations using finite element models, it originates from the numerical errors (discussed in Sects. 4.2.3, 4.2.4, and 4.2.6) and is most likely of an epistemic nature. For example, in structural dynamics problems, when explicit integration methods are used to solve Eq. (4.22), the noise is the high-frequency components of the calculated response, which are associated primarily with numerical modeling errors rather than with any physical processes in the analyzed object itself. Implicit integration methods (see Sect. 4.2.3) have a damping property that suppresses high-frequency vibrations (Bathe and Wilson, 1976). In any case, both in experiments and in computer simulations, digital filtering procedures are applicable to eliminate noise in the dynamic response of the system (Rabiner and Gold, 1975; Smith, 1999).

The best way to validate the mathematical models used and the computer calculation results obtained with them is to conduct an appropriate experiment. However, experiments are not always feasible – in particular, under terrestrial conditions, due to gravity, it is practically impossible to test large space structures, e.g., thin-film space reflectors (see Figure 3.1). Therefore, during design, it is sometimes necessary to substantiate calculation results solely through analytical methods, without relying on experimental data.

When using FEM for structural modeling, one way to analyze the accuracy of the constructed model is to consider several mesh variants with successively smaller finite element sizes (and, correspondingly, an increase in the number of elements and the number of degrees of freedom in the model) and compare the responses calculated on different meshes. For elliptic type problems, the errors in the obtained values of the response variables follow Eq. (4.38), i.e., they decrease according to a power law with a certain power value. For other types of problems, the dependence of the error on the number of degrees of freedom in the model may differ from this law by the exponent, but the general nature of the dependence – error reduction with finite element mesh refinement – remains the same. This nature of the dependence persists even for highly nonlinear dynamic problems associated with modeling the structures of various vehicles in crashes (Spirochkin, 2023).

Another way to validate calculation results in the absence of experimental confirmation is to perform parallel calculations using fundamentally different methods and models. For example, in structural dynamics problems, as an alternative to direct integration of Eq. (4.22) obtained by FEM, one can use quasi-static analysis with dynamic amplification factors, modal analysis or other analytical methods.

The use of analytical methods for describing human factors was mentioned in Sect. 3.2. A number of mathematical models have been developed for analyzing human behavior, however, the level of development of these models and simulation methods cannot be considered sufficient for their use in the design of technical systems, taking into account all manifestations of these factors (Spirochkin, 2023). In particular, methods for predicting human behavior under conditions of malicious actions – both attackers and potential victims – are analytically weak. A kind of verbal model of an attacker can be found, for example, in the part of the US Code of Federal Regulations devoted to the development of physical protection systems for nuclear facilities, including protection against "radiological sabotage", i.e., "a determined violent external assault, attack by stealth, or deceptive actions, including diversionary actions, by an adversary force". Adversary force can consist of one or more individuals, groups or their combination. These can be "well-trained (including military training and skills) and dedicated individuals, willing to kill or be killed, with sufficient knowledge to identify specific equipment or locations necessary for a successful attack" (CFR Part 73, § 73.1).

A mathematical description of similar malicious behavior, if such a description is possible at all, should obviously employ agent-based modeling methods (Spirochkin, 2023). However, at present, when an appropriate methodology that meets design goals is undeveloped, other methods must be used to address epistemic uncertainty inherent in marginal manifestations of human factors. One possible approach is discussed in Sect. 8.3.

4.4.2 Applicability of the Analytical Approach to Uncertainty Reduction

The analytical methods presented in this chapter are applicable to the control, including reduction, of uncertainty generated by sources 2 – 6 listed in Table 3.1, i.e., when the uncertainty contains an epistemic component. Reduction of uncertainty is achieved through the construction of adequate mathematical models of objects of analysis used to substantiate design decisions through calculations. Analytical methods are also applied to process the results of experiments conducted for the same purpose, as well as in a number of other design situations characterized by uncertainty, contributing to its reduction.

As shown above, the correct application of analytical methods allows designers to control and limit the errors that occur in calculations. Furthermore, the propagation of errors associated with uncertainty in input data regarding the properties of the designed system and the loads acting on it during operation, which are not known with sufficient accuracy, can be controlled through calculations. However, analytical methods are not suitable for controlling the occurrence of errors in input data, which are encompassed by the concept of aleatory uncertainty.

Analytical methods play an important role in the mathematical reconstruction of objects with which designers have to deal. Reconstruction problems can be characterized by significant uncertainty, primarily epistemic. The current level of development of analytical methods and mathematical models that can be applied at the design stage remains insufficient for managing uncertainty associated with human factors.

References

Abramowicz, W. (2001) Macro element method in crashworthiness of vehicles. In J.A.C. Ambrósio (ed.) *Crashworthiness. Energy management and occupant protection*, Part II. Springer, Wien, New York, pp. 83–138.

Abramson, H.N. (ed.) (1966) The dynamic behavior of liquids in moving containers with applications to space vehicle technology. NASA SP-106, NASA, Washington, DC.

Antony, J. (2014) *Design of experiments for engineers and scientists*. 2nd edn. Elsevier, London.

Bathe, K.-J. and Wilson, E.L. (1976) *Numerical methods in finite element analysis*. Prentice-Hall, Englewood Cliffs, NJ.

Belytschko, T., Krongauz, Y., Organ, D., Fleming, M. and Krysl, P. (1996) Meshless methods: an overview and recent developments. *Computer Methods in Applied Mechanics and Engineering*, 139(1-4):3–47.

Cabanes, M. and Späth, B. (2024) The McKay Conjecture on character degrees. arXiv > math > arXiv:2410.20392. (Submitted on October 27, 2024). https://arxiv.org/abs/2410.20392. Assessed December 19, 2025,

CFR Part 73. Code of Federal Regulations/Title 10/Chapter I/Part 73 – Physical Protection of Plants and Materials. https://www.ecfr.gov/current/title-10/chapter-I/part-73. Accessed December 21, 2025.

Chen, S.S. and Rosenberg, G.S. (1975) Dynamics of a coupled shell-fluid system. *Nuclear Engineering and Design*, 32(3):302–310.

Chen, J.-S., Hillman, M., and Chi, S.-W. (2017) Meshfree methods: progress made after 20 years. *Journal of Engineering Mechanics*, 143(4):04017001.

Dodge, F.T. (2000) *The new "dynamic behavior of liquids in moving containers"*. Southwest Research Institute, San Antonio, TX.

Dowell, E.H. (2015) *A modern course in aeroelasticity*. Fifth revised and enlarged edition. Springer, Switzerland (Solid Mechanics and Its Applications 217).

Du Bois, P., Chou, C.C., Fileta, B.B., Khalil, T.B., King, A.I., Mahmood, H.F., Mertz, H.J. and Wismans, J. (2000) *Vehicle crashworthiness and occupant protection*. (Editors: P. Prasad, J.E. Belwafa). American Iron and Steel Institute, Southfield, MI.

Durakovic, B. (2017) Design of experiments application, concepts, examples: state of the art. *Periodicals of Engineering and Natural Sciences*, 5(3):421–439.

Faddeev, D.K. and Faddeeva, V.N. (2002) Vychislitelnye metody lineinoi algebry (*Computational methods of linear algebra*). 3rd edn., reprinted. Lan, Saint Petersburg (in Russian).

Fisher, R.A. (1971) *The design of experiments*. 9th edn. Hafner Press, New York.

Fried, I. (1972) Perturbation errors in the finite-element method. *Journal of Applied Mechanics*, 39(2):629–631.

Huerta, A., Belytschko, T., Fernández-Méndez, S. and Rabczuk, T. (2004) Meshfree methods. In *Encyclopedia of Computational Mechanics*. Edited by E. Stein, R. de Borst and T.J.R. Hughes. John Wiley & Sons, Part I, Chapter 10.

Ibrahim, R.A. (2005) *Liquid sloshing dynamics. Theory and applications*. Cambridge University Press, Cambridge, UK.

Levin, V.A. and Vershinin, A.V. (2015) Chislennye metody. Parallelnye vychisleniya na EVM (*Numerical methods. Parallel computing*). Fizmatlit, Moscow (Nonlinear computational mechanics of strength, in 5 volumes, Volume II) (in Russian).

Mavromanolakis, G., Manousos, T., Kechri, M., Kollia, P.L. and Kanellopoulos, G. (2019) Studying, designing and 3d-printing an operational model of the Antikythera Mechanism. *Open Schools Journal for Open Science*, 1(3):70–77.

McKay, J. (1972) Irreducible representations of odd degree. *Journal of Algebra*, 20(2):416–418.

Mikishev, G.N. and Rabinovich, B.I. (1971) Dinamika tonkostennykh konstruktsii s otsekami, soderzhaschimi zhidkost (*Dynamics of thin-walled structures with compartments containing liquid*). Mashinostroenie, Moscow (in Russian).

Montgomery, D.C. (2013) *Design and analysis of experiments*. 8th edn. John Wiley & Sons, New York.

Oberkampf, W.L., Trucano, T.G. and Hirsch, C. (2003) Verification, validation, and predictive capability in computational engineering and physics. Report SAND2003-3769. Sandia National Laboratories, Albuquerque, NM, and Livermore, CA.

Oberkampf, W.L. and Roy, C.J. (2010) *Verification and validation in scientific computing*. Cambridge University Press, Cambridge, UK.

Rabiner, L.R. and Gold, B. (1975) *Theory and application of digital signal processing*. Prentice-Hall, Englewood Cliffs, NJ.

Roy, C.J. and Oberkampf, W.L. (2011) A comprehensive framework for verification, validation, and uncertainty quantification in scientific computing. *Computer Methods in Applied Mechanics and Engineering*, 200(25–28):2131–2144.

Smith, S.W. (1999) *The scientist and engineer's guide to digital signal processing*. 2nd edn. California Technical Publishing, San Diego, CA.

Spirochkin, Y.K. (1987) Metodika raschyota dinamicheskogo nagruzheniya gruzhov pri avariinoi posadke letatelnogo apparata (Method for analyzing dynamic loads on items of mass during an aircraft emergency landing). Dissertation. NPO *Energia*, Kaliningrad (now Korolyov), Moscow region (in Russian).

Spirochkin, Y.K. (1993) Computer simulation of structural dynamics in accidents. *Mathematical Models and Computer Simulations*, 5(6):85–103 (in Russian).

Spirochkin, Y.K. (1994) Special finite elements for structural dynamics problems. *Mathematical Models and Computer Simulations*, 6(8):85-91 (in Russian).

Spirochkin, Y.K. (1997) System of vibration-noise diagnostics of reactor installations of the WWER type. Software for finite element analysis of structures: *NewTone*. Version 1.5. Special configuration for calculation of dynamic characteristics of structure in liquid. User's manual. Concern 'Rosenergoatom'. Center for Non-destructive Testing and Diagnostics: DIAPROM, Moscow (in Russian).

Spirochkin, Y.K. (1998) Konechnoelementnoe modelirovanie dinamiki obolochek, vzaimodeistvuyuschikh s tonkimi sloyami zhidkosti (Finite element modeling and simulation of the dynamics of shells interacting with thin layers of liquid). Prikladnye problemy prochnosti i plastichnosti. Chislennoe modelirovanie fiziko-mekhanicheskikh protsessov (*Applied problems of strength and plasticity. Numerical modeling and simulation of physical and mechanical processes*). Interuniversity collection, Issue 58, pp. 110-121 (in Russian).

Spirochkin, Y. (2011) Special finite elements in structural dynamics. *Vibration Problems ICOVP 2011. Supplement. The 10th International Conference on Vibration Problems*. Technical University of Liberec, Liberec, Czech Republic, pp. 168–173.

Spirochkin, Y. (2023) *Human factors and design*. Springer, Singapore.

Strang, G. and Fix, G.J. (1973) *An analysis of the finite element method*. Prentice-Hall, Englewood Cliffs, NJ.

Tang, Z., Liu, F.J., Guo, S.H., Chang, J. and Zhang, J.J. (2016) Evaluation of coupled finite element/ meshfree method for a robust full-scale crashworthiness simulation of railway vehicles. *Advances in Mechanical Engineering*, 8(4):1–13.

Tikhonov, A.N. and Samarskii, A.A. (1977) Uravneniya matematicheskoi fiziki (*Equations of mathematical physics*). 5th edn. Nauka, Moscow (in Russian).

Timoshenko, S.P. and Young, D.H. (1974) *Vibration problems in engineering*. 4th edn. John Wiley & Sons, New York.

Volmir, A.S. (1967) Ustoichivost deformiruemykh system (*Stability of deformable systems*). 2nd edn,, revised and enlarged. Nauka, Moscow (in Russian).

Volmir, A.S. (1976) Obolochki v potoke zhidlosti i gaza. Zadachi aerouprugosti (*Shells in liquid and gas flows. Aeroelasticity problems*). Nauka, Moscow (in Russian).

Volmir, A.S. (1979) Obolochki v potoke zhidlosti i gaza. Zadachi gidrouprugosti (*Shells in liquid and gas flows. Hydroelasticity problems*). Nauka, Moscow (in Russian).

Wright, J.R. and Cooper, J.E. (2007) *Introduction to aircraft aeroelasticity and loads*. John Wiley & Sons, Chichester, UK.

Zienkiewicz, O.C. (1971) *The finite element method in engineering science*. McGraw-Hill, London.

Zienkiewicz, O.C. and Morgan, K. (1983) *Finite elements and approximation*. John Wiley & Sons, New York (Wiley-Interscience Publication).

Zukas, J.A., Nicholas, T., Swift, H.F., Greszczuk, L.B. and Curran, D.R. (1982) *Impact dynamics*. John Wiley & Sons, New York.

5 Reducing Uncertainty Using New Experimental Data

In this chapter, the approaches to reducing uncertainty are described that are based on obtaining new experimental data. These include, first and foremost, such obvious measures as the refinement of probabilistic characteristics approximating the stochastic properties of a technical system and load parameters recorded in research experiments, and the determination of the actual properties of individual system specimens and loading conditions during certification or other tests. Attention is paid to an approach that uses the Bayesian method to refine a priori probability estimates based on newly acquired data. This method, well known to researchers, has not yet gained widespread application in design. Two other design approaches, which also involve obtaining experimental data, are much more familiar to designers: pilot projects and in-service monitoring and diagnostics. For each approach considered, its potential for minimizing uncertainty is analyzed and its inherent limitations are discussed.

5.1 Refinement of Probabilistic Characteristics

5.1.1 The Role of Experimental Refinement and the Problems It Should Solve

Uncertainty in design can be reduced by refining the probabilistic characteristics of factors that influence the parameters of the system being created and that have a stochastic nature. Such refinement is especially necessary when the system differs significantly from prototypes in terms of operating conditions (environmental factors, operating modes, designated service life), design features (configuration, dimensions of the main elements and their connections), materials used and manufacturing technologies. In this case, it should be expected that the stochastic variation of geometric characteristics of elements, properties of materials and connections, as well as load parameters can go beyond the statistical data collected for prototypes.

The problems of experimental refinement of probabilistic characteristics, according to the provisions of Sects. 3.2 and 3.3, include:

1) determination of distribution laws, the mean values and variances (or standard deviations) of the properties of new materials and geometric parameters of structural elements of unusual configuration, as well as materials and elements manufactured using new technologies, or direct acquisition of data

DOI: 10.1201/9781003670575-5

on the stochastic variation of the bearing capacity of such elements before the start of operation of the system;

2) substantiation of forecasts for changes in the probabilistic characteristics of material properties, geometric parameters and load-bearing capacity of structural elements during their service life under the influence of adverse operational factors;
3) substantiation of forecasts for changes in the probabilistic characteristics of maximum loads acting on the system over a long life cycle, including loads caused by natural impacts;
4) refinement of distribution laws, the mean values and variances of maximum load parameters, taking into account the possible "splitting" of statistical data into two sets: the main, predominant in volume, and an additional, minor in volume, extreme set, which can be key to substantiating the strength of the system structure, taking into account rare adverse events.

Solving all of these problems contributes to reducing the aleatory uncertainty generated by source 1 (Table 3.1). Furthermore, solving the second, third, and fourth problems can reduce the epistemic uncertainty generated by sources 4, 6, and 7.

5.1.2 Experimental Refinement Methodology

Experimental refinement is achieved by obtaining new data in the form of statistical series, specific for the system being created, and processing this data to determine reasonable probabilistic approximations. These approximations should ensure reduced error when estimating the state parameters and behavior of the designed system in predicted operating modes.

New experimental data can be obtained during tests related to the category of *phenomena-exploration experiments* (Oberkampf et al., 2003). These include, in particular, tests to determine the properties of basic materials and joints (mechanical, welded and others), which are carried out in laboratory conditions on samples and structural elements. Manufacturing experimental objects, conducting tests and processing their results can be one of the intermediate stages of design, which must precede the adoption of final design decisions. To obtain the statistical characteristics of the results of serial tests, methods described in the publications on design of experiments cited at the end of Sect. 4.3.2 are applicable. There are well-developed recommendations for accounting for aleatory uncertainty in the data of mechanical tests – see, for example, Stepnov (1985), and also those relating to a wider range of experiments (ASME PTC 19.1-2005). If such tests reveal increased dispersion of properties compared to prototypes, then it may be necessary to improve technological processes, strengthen structural elements, or increase the factors of safety.

Similar experiments can provide data on the stochastic variations in the geometry of elements and joints. The main share in such variations is made up of initial technological imperfections: thinning and deflection of the walls of shell elements, eccentricities and other deviations from the design forms and sizes, which are often combined with residual stresses after various technological operations.

Statistical data on operational loads for a new system can be obtained in research facilities using small-scale physical models or full-size mock-ups, on prototypes of the system manufactured before the start of its serial production, as well as on specimens already put into operation, equipped with appropriate recording devices. The degree of reproduction of real processes using experimental objects of the first two types in research facilities is determined by compliance with the similarity criteria. Full-scale experiments on prototypes have no limitations in terms of similarity, but they are usually expensive and can only be carried out when the system design has been completed and the manufacturing stage has begun. If the results of these experiments require any changes to the system, this could become a problem due to additional costs and time constraints. The data recorded during the operation of a serial specimen of the system over a long period of time is the most valuable information resource. This resource can contain records of relatively rare natural impacts, but its technical support is associated with costs and its applicability is primarily aimed at future projects. However, since the parameters characterizing natural impacts typically depend weakly on the properties of the system (primarily due to the large scale of natural phenomena), they can be obtained using similar objects that have been in operation for a long time (and, therefore, can act as prototypes). An example is atmospheric turbulence data collected using aircraft of various types (Taylor, 1965).

Control of errors in probability estimates obtained according to this methodology can be carried out by analyzing their sensitivity to the variability of the volume of the statistical base and the type of distribution law (if it is possible to consider distributions of different types).

Along with its undoubted advantages, the described approach has a significant shortcoming: its application determines not the actual properties of the elements of each specific specimen of the system under development or the actual values of the loads that will act on them during operation, but rather the parameters of probabilistic approximations of the corresponding statistical populations that apply to all specimens of the system. This approach can reduce uncertainty for the entire fleet of serial-produced system specimens and a range of typical operating conditions, varying within predictable limits corresponding to reasonable probabilities. However, it is impossible to guarantee that a specific specimen has precisely known properties or that it will not be subjected to loads exceeding probabilistic estimates throughout its service life.

5.2 Determination of the Actual Properties of Individual Specimens of the System and Loading Conditions

The actual properties of a specific system specimen, including the characteristics of static strength, vibration resistance, or impact resistance, etc., can be obtained through the performance check experiments. These experiments fall under the category of *system and certification tests* (Oberkampf et al., 2003). Determination of actual properties is also possible in experiments to identify and refine mathematical models, known as *mathematical-model-development experiments* and *calibration* (*model-updating*) *experiments* (ibid). The data acquired in this way are often applicable to assessing the properties of a real system beyond the immediate objectives of the experiment. For example, the results of horizontal frequency tests[1] of the Soviet space shuttle *Buran* (see Sects. 6.4.2, 7.1.1 and 7.4.1) – natural frequencies and vibration mode shapes of the structure, which are necessary primarily to substantiate safe operation taking into account aeroelasticity and other dynamic effects, – were used to validate its finite element model constructed to analyze the behavior of the structure in an emergency landing (Spirochkin 1987, 1993).

The actual maximum values of loads and their time profiles can be determined in phenomena-exploration experiments mentioned in Sect. 5.1.2. In analyzing the emergency landing of *Buran* (see Sect. 7.4.1), one of the problems was determining the loads acting on the fuselage from the landing surface. For the case of emergency landing on the ground, these loads were calculated using experimental data accumulated during landing tests of Soviet reentry ballistic-type capsules. To determine the interaction of the lower part of the *Buran* fuselage with the landing surface under conditions of a glancing impact at a speed characteristic of an airplane-type landing, a special experimental facility was proposed (Spirochkin, 1988), subsequently patented (Patrushev et al., 1989).

The described approach eliminates the shortcoming of probabilistic assessment noted in Sect. 5.1.2 – averaging data over a statistical population – and can provide a reduction in aleatory uncertainty as applied to a specific specimen of the system (within the limits related mainly to the first source of uncertainty presented in Table 3.1). However, it has its inherent problems:

1) significant costs for conducting experiments for each specimen;
2) obtaining only its initial properties – information about their deterioration during operation remains unknown, and
3) lack of information on actual operational loads acting on each specimen.

To solve these problems, other methods are obviously needed.

5.3 Bayesian Approach: Power, Advantages, and Pitfalls

5.3.1 On the Applicability of the Bayesian Method to Design Problems

The Bayesian method, which allows subjective probability estimates to be refined as new experimental data becomes available, appears to be a useful tool for reducing the uncertainty associated with rare events for which little or no statistical data are available (see Sect. 2.2.2). The hypothesis analyzed using Eqs. (2.12) and

[1] Vibration tests with the test object in a horizontal position.

(2.13) may relate to the type of distribution law of a random variable describing some property of an object, to the probabilistic characteristics of this property or process, as well as the probability of an event of interest to designers.

With a sufficiently large likelihood ratio, determined by Eq. (2.13), it becomes possible to substantiate a design decision made under uncertainty. Decision-making is most challenging when newly acquired information does not provide direct confirmation of the working hypothesis or its alternative, but is only indirect evidence. By indirect evidence, we mean data characterizing not the object of interest itself, e.g., an event in question, but the observed circumstances that testify in favor of its implementation. Indirect, or circumstantial, evidence is quite likely for rare events.

Taking such evidence into account presupposes knowledge of the relationship between the observed circumstance and the object of analysis – the event, property, or process. This, in turn, implies a virtual construction of the object – at least in the mind of the decision maker. Thus, decision-making under uncertainty based on circumstantial evidence is akin to the analytical reconstruction of an object based on available fragments of information (Sect. 4.3). The use of the Bayesian method, which combines experimental data (newly obtained evidence) and analytical refinement, in this context allows it to take its rightful place among the methods of analytical reconstruction. The Bayesian method has proven useful in many areas of research, but has not yet found widespread application in the design of technical systems. Perhaps the main reason for this is that it works well in the presence of reliable prior probabilities, which are often lacking in design problems, whereas its application with unfounded prior probabilities can lead to inconclusive or even incorrect results (Persi and Freedman, 1986; Taleb, 2020).

Before examining the applicability of the Bayesian approach to design, let us turn to two historical examples that demonstrate the importance of taking into account indirect evidence in uncertain situations associated with design activities.

5.3.2 The Importance of Indirect Evidence in Uncertain Situations: Historical Examples

The first example comes from an initiative by Soviet nuclear physicist G.N. Flyorov[2], dating back to 1942. While reading foreign physics journals in the university library of Voronezh[3], he discovered that articles on uranium fission and related problems had disappeared from the latest issues (previously such articles were published regularly). Flyorov concluded that foreign research on nuclear chain reaction had become classified, as prospects for military applications had

[2]Flyorov, Georgii Nikolaevich (1913-1990), worked on the Soviet atomic project from 1943 to 1954, then he was involved in the synthesis of transuranic elements; one of the founding fathers of the Joint Institute for Nuclear Research in Dubna, academician of the USSR Academy of Sciences.

[3]Voronezh is a city in the central part of European Russia, a regional industrial and scientific center.

presumably emerged. He wrote about this to I.V. Kurchatov[4], to the Science Committee of the State Defense Committee, and to the Soviet leader Stalin himself, urging the continuation of stalled domestic research. Flyorov's appeals played a decisive role in organizing Soviet R&D to create an atomic bomb (NRCKI, 1998). Subsequent events confirmed the correctness of Flyorov's assumption: German physicists began developing atomic weapons first, and then American scientists followed their way. However, at the analysis stage, it was the disappearance of the relevant information from publicly available sources that served as indirect evidence in favor of this assumption.

A second example is the decision made by S.P. Korolyov, Chief Designer and head of the Soviet lunar project, regarding the state of the lunar surface, as described in Sect. 2.4.3. The decision was made in the absence of precise data and amidst disagreements among planetary scientists. While acknowledging the Chief Designer's intuition and his determination to assume responsibility in the face of uncertainty, it should nevertheless be noted that there was circumstantial evidence in support of his position – he referred to the opinion of the "majority of scientists" (Gallai, 1985).

In both cases described, the Bayesian method was not used. It was unlikely to have been known to engineers at the time. Even if it had been applied, it would hardly have been effective, since it was impossible to take any a priori probability estimate.

5.3.3 Applying the Bayesian Approach to the Challenger Disaster

The Bayesian approach is applicable to testing hypotheses about the development of uncertain critical situations, one of the possible outcomes of which is an adverse event – an accident or catastrophe. Predicting such a development and substantiating the forecast is of significant interest for safety analysis. Below, we demonstrate the feasibility of forecasting using the example of the disaster that occurred with the space shuttle *Challenger* during launch from Cape Canaveral on January 28, 1986. To the author's knowledge, Bayesian probability estimation of such an event was not conducted at the time, so this example is for illustrative purposes only.

The launch took place in subzero weather, with the air temperature not exceeding −1°C. Engineers at *Thiokol*, the manufacturer of the *Space Shuttle* solid rocket boosters, and at NASA's Marshall Space Flight Center were aware of the negative impact cold temperatures had on the reliability of the rubber rings that sealed the booster sections' joints. If the seals lost their elasticity, they would no longer provide the tightness, and combustion gases can escape through the joints. For these reasons, engineers urged the mission to be rescheduled for warmer

[4]Kurchatov, Igor Vasilievich (1903-1960), was a Soviet physicist, the chief scientific director of the atomic project in the USSR; one of the initiators of the use of nuclear energy for peaceful purposes; the founder and first director of the Institute of Atomic Energy, now the Russian National Research Center *Kurchatov Institute*, Moscow.

weather, but NASA management ignored them. The decision was made to proceed with the launch despite the uncertainty of mission success given the unreliability of the seals (Presidential Commission, 1986; Committee on Science and Technology, 1986). Managers considered this uncertainty to be within the acceptable risk of a vehicle-loss accident, corresponding to a probability of approximately 10^{-5} (Feynman, 1986).

The engineers' bad premonitions were confirmed: analysis of the launch footage showed smoke coming from the joint between sections of the right booster, near where it attaches to the external fuel tank. A leak of hot gases (with a temperature of approximately 2760°C) led to the almost complete evaporation of the sealing ring, but a plug of aluminum oxides formed during fuel combustion temporarily sealed the leak. Between 37 and 60 seconds into the flight, aerodynamic loads and vibrations caused this plug to pop out, sending a stream of hot gases burning through the booster's steel wall and striking the external fuel tank. A few seconds later, liquid hydrogen began leaking from the tank. At 73 seconds into the flight, the right booster's lower attachment to the external fuel tank failed, causing the booster to rotate in its upper attachment and penetrate the external fuel tank's shell near the liquid oxygen tank. When the released propellant components mixed with each other, an explosion occurred. The disaster resulted in the deaths of all seven crew members, the loss of the space shuttle, total losses of $8 billion, and the grounding of the *Space Shuttle* program for 32 months.

Let us estimate the subjective probability of a catastrophic outcome of an uncertain pre-launch situation using the Bayesian method. As hypothesis, we will consider mission failure due to a leak in the joints of the solid rocket booster sections. We will consider evidence to be the low ambient temperature observed in the launch area. Note that this evidence regarding a leak in the booster sections' joints (and, consequently, mission failure) is indirect. An a priori estimate of the mission failure probability can be taken based on the known failure rate of solid-fuel rockets: $P(H) = 0.02$ (Feynman, 1986). The probability of evidence A occurring, when hypothesis H is true, can be characterized as average: $P(A|H) = 0.5$. The probability of the realization of evidence A without considering H can be estimated according to statistical data on subzero temperatures observed in the Cape Canaveral area at the end of January (NOAA, 2023): $P(A|\mathit{not}\ H)$ does not exceed 0.25. Substituting these values into Eq. (2.13), we obtain the likelihood ratio: $L = 1.96$.

The obtained value of L leads to a nearly twofold increase in the posterior estimate of the probability of mission failure due to the thermal unreliability of the rubber rings of the solid rocket boosters compared to the a priori estimate: according to Eq. (2.12), $P(H|A) = 0.0392$ (versus $P(H) = 0.02$). If *Challenger* were a civil aircraft, the resulting probability would have to be considered high, as it characterizes *frequent* events (AP-25, 2009), and the associated risk would have to be considered unacceptable, requiring immediate mitigation measures (ICAO, 2018). However, no such measures were taken.

This example demonstrates not only the possibility of substantiating the subjective probability of an event based on indirect evidence to reduce the uncertainty of its occurrence, but also the danger of ignoring such evidence if it points to a likely adverse outcome. At the same time, it should be noted that the likelihood ratio in favor of the adverse outcome hypothesis, obtained on the basis of the available input data, was not so convincing as to justify its unconditional acceptance. Even if such an analysis had been conducted before the ill-fated launch, a likelihood ratio of less than 2 would hardly have been considered convincing. Strong evidence supporting the hypothesis being analyzed is considered to be one yielding a likelihood ratio of at least 10. Thus, the fate of the *Challenger* crew was, alas, sealed.

5.3.4 Bayesian Risk Assessment of Orbital Debris Falls on Earth

As the frequency of rocket launches and the number of objects in low Earth orbit increases, the risk of their uncontrolled re-entry into the atmosphere and impact on the Earth's surface becomes a serious problem that concerns the scientific community, relevant international organizations, and the public (OSI, 2023; IADC, 2025; Lewallen, 2025). This problem is exacerbated due to the expansion of the private space business and the development of technologies for returning rocket parts to Earth for reuse. The core of the problem lies in the uncertainty of the risks associated with the increasingly frequent re-entries of various objects. These re-entries can range from controlled descents, when the object remains functional and follows its intended trajectory, to chaotic falls, typical of abandoned rocket bodies, failed satellites, their parts, other artifacts "forgotten" in orbit, and structural fragments of different sizes resulting from collisions of these objects or other destructive processes. Inoperative and uncontrolled spacecraft components, their fragments, and other similar objects constitute the category of *orbital debris*.

According to information presented in the quarterly publication of the NASA Orbital Debris Program Office (NASA, 2025), there were 29,837 artificial objects orbiting Earth on June 4, 2025 (269 more than in the previous quarter). This number includes active and decommissioned spacecraft, spent rocket bodies and other orbital debris cataloged by the US Space Surveillance Network. The risk of an uncontrolled re-entry of some of these objects exists, and intuition suggests that it should increase over time. If the re-entering object is large enough, between 10 and 40% of its mass could survive the high temperatures as it moves through the atmosphere and can reach the Earth's surface (Ailor and Patera, 2007), posing a potential threat to people, civil infrastructure, industrial facilities, and vehicles, including aircraft in flight. However, there are no reliable estimates of such risk, and the problem under consideration combines aleatory and epistemic uncertainty.

Attempts to study the risk associated with space objects falling to Earth have been underway for a long time. One of the first was an analysis of the safety of the territory of the Russian Federation carried out by a company headed by the author

(Dynamika, 1992). This analysis was commissioned by the State Committee of the Russian Federation for Emergency Situations, the predecessor of the current Ministry of Emergency Situations. The resulting estimate of the probability of an uncontrolled fall of an artificial Earth satellite was 1×10^{-7} 1/year. Based on this estimate, such events can be considered to be on the border between low-probability and extremely unlikely (AP-25, 2009). The obtained value corresponds to the target safety indicator of the probability of a catastrophe for any reason. It was determined theoretically and had no experimental confirmation.

Another study in this area was an analysis of the risk of collisions between falling space debris and commercial aircraft flying over the United States, departing abroad, or arriving from other countries (Patera, 2008). The object chosen for probability assessment in this study (a specific aircraft fleet) differed from that used in the Russian analysis (territory), and the probability of a collision between any aircraft and falling space debris was estimated to be 0.769×10^{-8}. This estimate, like the Russian one, was not supported by any experimental data.

Over the years, evidence has emerged of uncontrolled falls of orbital debris. To date, three such incidents have been documented:

- in 1997, a person was hit by a small fragment of a Delta II rocket (Ailor and Patera, 2007);
- in 2021, a Falcon 9 pressure vessel landed on a farm in Washington state (Roulette, 2021);
- in 2024, a piece of hardware released from the International Space Station punched through a home in Naples, Florida (Garcia, 2024).

The availability of such evidence (all of which is direct) allows us to refine the previously obtained a priori estimate of the risk in question using the Bayesian approach. The hypothesis being tested is the uncontrolled fall of orbital debris to the Earth. As an a priori estimate of the probability of such an event, we take our theoretically determined value: $P(H) = 1 \times 10^{-7}$ 1/year (Dynamika, 1992). All three incidents that occurred within 33 years after obtaining this estimate can be considered as cumulative evidence A. The conditional probability of evidence A, given the truth of hypothesis H is equal to: $P(A \mid H) = 1$ (1/year), since all these events occurred precisely as a realization of this hypothesis. The probability that evidence A is not related to hypothesis H, i.e. the falls of objects occurred independently of it for completely random reasons, can be estimated by the frequency of these events: $P(A \mid not\ H) = 3/33 = 0.0909$ (1/year). Substituting the values of $P(H)$, $P(A \mid H)$, and $P(A \mid not\ H)$ into Eq. (2.13) yields the likelihood ratio: $L = 11$.

This value is high enough to confirm the validity of the tested hypothesis. Based on the found value of L, the refined, a posteriori estimate of the probability of an uncontrolled fall of orbital debris to Earth, according to Eq. (2.12), is $P(H|A) = 1.1 \times 10^{-6}$. This estimate "shifts" the events under consideration from extremely unlikely to low-probability ones (AP-25, 2009). It exceeds, albeit slightly, the accident probability assumed in the design of technical systems

(10^{-6}). With the acquisition of new evidence on uncontrolled falls of orbital debris, updated a posteriori estimates of the probability of such events can be obtained, and these estimates will only increase.

In October 2025, a smoking fragment of carbon fiber structure, believed to be part of a rocket, was discovered in Western Australia (David, 2025). At the time of writing these lines, the Australian Space Agency has not yet completed its investigation into its origin. However, if this incident serves as further evidence of an uncontrolled fall of orbital debris, the Bayesian a posteriori estimate of the probability in question increases to 0.932×10^{-5}.

The international community currently recommends design and organizational measures aimed at reducing the formation of orbital debris and preventing its uncontrolled return to Earth (OSI, 2023; IADC, 2025). These recommendations are followed and supported by national standards, particularly those of Russia and the USA. If a spacecraft or a rocket part is to be disposed of by re-entry, then, given the possibility of an uncontrolled fall, fragments capable of reaching the Earth's surface should not pose an excessive risk to people or property. A value of 10^{-4} is recommended as the upper limit of the probability corresponding to the expected human casualties. The Bayesian probability estimate obtained above is still far from this limit. Furthermore, there are no known cases of orbital debris falling on nuclear power plants or other safety-critical ground infrastructure, nor of collisions with aircraft in flight. Therefore, it is difficult to imagine how to respond to such low probabilities. However, the obvious trend of their increase over time makes one think about preventative measures that designers will need to take – perhaps in the near future – to avoid disasters caused by possible falls.

5.3.5 Caution when Using in Design

Thus, the examples described in Sects. 5.3.3 and 5.3.4 demonstrate the potential of the Bayesian method for reducing uncertainty in design. This uncertainty is primarily aleatory, characteristic of random events that are possible during the operation stage. It is associated with sources 1, 2, and 4, presented in Table 3.1. In addition, an important feature is noted: some of the evidence on which the desired probability estimates are based may be indirect. This feature does not limit the applicability of the method and even enhances its capabilities in assessing the risk associated with rare events for which there are no and cannot be reliable statistics. The refinement of the probability estimate on the basis of new evidence obtained experimentally can, if this evidence is circumstantial, be combined with an analytical reconstruction of the object of direct analysis based on known cause-and-effect relationships.

However, along with these advantages, which determine the power and versatility of the Bayesian approach, it also has its drawbacks. The first of them, noted in Sect. 5.3.1, is the potential unreliability of prior probabilities, which is to be expected for rare events, as well as for objects that researchers and designers encounter for the first time and whose characteristics are unknown. A posteriori estimates obtained on such a shaky basis using one or two pieces of new evidence, although formally capable of providing a more accurate understanding of the object of analysis, can also lead away from the real state.

The second drawback, noted in Sect. 5.3.3, is that the likelihood ratios obtained using the Bayesian method are only trustworthy when they are significantly greater than 1 and closer to 10. Values less than 5 obviously cannot be considered convincing enough to accept the hypothesis being tested.

Another drawback, inherent not so much to the Bayesian method itself as to its use by inexperienced analysts, can be the limited set of testable hypotheses. The preference for one hypothesis over another, resulting from the formal application of this method, generally speaking, does not indicate its correctness relative to reality, which can be much more diverse than a dichotomy.

There are a number of other mathematical and philosophical aspects of the Bayesian approach that are not so obvious, but being a kind of pitfalls, they can limit its applicability in practical, including engineering, problems (Wang, 2004; Gelman, 2008; Jones, 2021).

5.4 Pilot Projects

Innovations aimed at implementing scientific ideas in new technical systems may be characterized by high levels of epistemic uncertainty. To reduce it, the concept of a pilot project is used. This concept was mentioned in Sect. 2.4.1 as one of organizational ways to deal with uncertainty. Here we will look at it in more detail. Within this concept, neither the probabilities of various outcomes of innovation nor any other measurable indicators are assessed, but uncertainty as such is studied empirically. A pilot project may be a stage of a larger and long-term project (program) and follow any of the strategies below:

1) exploration of the space of possible outcomes through parallel promotion of several innovation options (by separate project teams) and identification of the most successful of them;
2) limiting the scope of innovation to an area corresponding to low (acceptable) risk, interpreted as the amount of potential losses, learning lessons and identifying corrective measures;
3) modeling key actions typical for implementing an idea in question, using a small representative example to test its feasibility, identify possible problems and find ways to solve them.

The first strategy was used by NASA in the 1960s when developing the lunar module. It was used in the first half of the 1990s by Japanese consumer electronics firms. Similarly, many automobile companies develop several concepts for a new car model, taking them to prototyping to fully understand the complex design space before selecting the best prototype for production (Pich et al., 2002).

The second strategy is largely inherent to innovations in defense and deterrence (Cocron and Aronhime, 2022). It is also applicable to other pressing problems with high potential for uncertain dangers, such as the integration of AI in aviation systems (Spirochkin, 2024).

The last strategy is familiar to the author from his participation in 2006-2008 in a program aimed at introducing a promising concept – *information support for the life cycle* (see Sect. 6.2.2) – in the design of transport-type nuclear reactors. The complexity, novelty, and degree of responsibility of this program necessitated a pilot project in the form of engineering work on a certain pipeline system as part of a specific reactor, with data exchange between the stages of work and the departments of the design organization performing them.

Direct "confrontation with uncertainty" in a pilot project allows designers to investigate such sources of uncertainty as the emergence of new properties (source 3 in Table 3.1), inaccuracies in the substantiations of design decisions (source 6), and unknown factors (source 7). Returning to the characterization of the pilot project concept as one of the organizational approaches to dealing with uncertainty, it should be noted that, unlike such ways as insurance (Sect. 2.4.2) or specialization (Sect. 2.4.3), it ensures a reduction in uncertainty.

5.5 In-service Monitoring and Diagnostics

5.5.1 The Essence of the Approach

Let us return to the idea of experimentally determining the actual properties of each specimen of a serially produced system and the loading parameters, which was discussed in Sect. 5.2. As one of the problems inherent in system and certification tests it was pointed out that they only allow the determination of initial properties, i.e. properties before the system is put into operation, and do not provide information on their deterioration during the service life. This problem, which limits the applicability of the described approach, can be addressed through *in-service monitoring and diagnostics*.

In-service monitoring includes observing the current state (health and operability) of the system during operation and, as well as operating conditions, by recording data characterizing this state and these conditions. Recording operating conditions creates the opportunity to solve another problem mentioned in Sect. 5.2 – obtain information on actual operational loads. Raw data are harvested through visual inspection and measuring instruments used by monitoring personnel, as well as sensors permanently installed at specific points in the system and working automatically. Diagnostics involves analyzing the collected data and assessing the system's state based on certain signs that may indicate possible malfunctions. The data that enable such an assessment to be carried out include both directly observable anomalies and defects, including those detected by measuring instruments (breakages, traces of wear or corrosion, macrocracks, etc.), and the parameters of the system's response to loads acting during normal operation or applied in special tests.

To assess the system's health based on its response, so-called *diagnostic models* are used. These are information objects that describe the relationship between certain features manifested in the system's response and the characteristics of its state that correspond to hidden faults. These features act as diagnostic signs, and the relationship is usually learned from previous experience. When the system is relatively simple, such a relationship may take the form of a small table listing diagnostic signs and the faults that correspond to them. For complex systems consisting of multiple elements, in which various working and accompanying processes occur, much more advanced mathematical objects can serve as diagnostic models: analytical dependencies, algorithms, graphs, expert systems, etc.

In-service monitoring and diagnostics operations, carried out periodically according to the established plan and after extreme events are or can serve as a source of data on the actual geometric parameters and properties of system elements changing over time, as well as the actual history of their loading.

The concept of in-service monitoring and diagnostics extends beyond the design stage and transfers some of the tasks of ensuring the safety and efficiency of the technical system being created to the operation stage. This represents a shift in the design paradigm, caused by existence of epistemic uncertainty and discussed in Chap. 2. Designers thereby acknowledge the impossibility of fully addressing the safety and efficiency problems based on a priori knowledge and equip the system with the means to obtain real data on events and processes during operation. This data enables the operator (the organization operating a system that exists as a unique specimen or a fleet of similar, serially produced systems with a long service life) to make informed decisions regarding system life cycle management.

The design organization that created the system typically has access to in-service monitoring data and is involved in the substantiation of these decisions. A *life cycle management based on actual state* offers an alternative to established plans for regular technical inspection, maintenance, repair, and replacement of elements. These plans, which are essentially a priori, may either be overly conservative, and then the alternative can improve economic indicators, or they do not sufficiently account for extreme operating conditions and accelerated element degradation – in this case, the alternative creates the potential for improved safety. The concept of life cycle management, which is a further development of the ideas of determining and taking into account the actual state of a technical system changing during its service life, is described in Sect. 6.2.

5.5.2 Applicability Across Industries

Currently, in-service monitoring and diagnostic techniques are implemented in various technical systems. Simple built-in diagnostic tools in the form of fault indicators can be found even in household appliances and mass-produced cars. The monitoring and diagnostics approach has become practically the standard in the design of expensive and safety-critical systems with long life cycles, such as civil infrastructure, industrial processing plants, NPPs, and so on.

Monitoring of the state of civil infrastructure and the loads these structures are subjected to during operation is a broad area of R&D due to the construction of large bridges, tunnels, high-rise buildings, etc. (Hejll, 2007; Chen and Ni, 2018; Sivasuriyan et al., 2021; Vardanega et al., 2022; He et al., 2022; Isac et al., 2025). For each specific site, data is collected on the degradation of concrete and steel structures under the influence of climatic and meteorological factors, as well as on operational, temperature, wind and seismic loads.

In industrial processing plants, monitoring of the state and diagnosis of hidden defects in the structures of pressure vessels, pipelines, and machine units is carried out, along with the detection of malfunctions and deviations of working processes from design parameters that could lead to an accident with production shutdown and other harmful consequences. Data characterizing temperature, pressure, composition of process gases and liquids and their flow rate, tightness of connections, power supply parameters, etc. are recorded. Monitoring and diagnostic tools and the communications they use can be integrated into automated process control systems (Chisholm, 1994; Smith, 2002; Ruilin et al., 2005).

In the nuclear energy industry, there are requirements for monitoring the working processes in a nuclear reactor and the state of the metal of its most important elements during operation. For example, according to the Russian regulatory framework, a NPP design must provide for (NP-001-15):

- "technical means and organizational measures aimed at preventing violations of the limits and conditions of safe operation";
- "technical means for monitoring the state" of the reactor and the NPP as a whole "under accident conditions, including severe accidents";
- "appliances and devices for confirming the operability of systems and elements (including devices located inside the reactor)", "monitoring the state of the metal (including welded joints) of equipment and pipelines", etc.

In order to apply the *leak before break* concept for safety substantiation, the NPP design must determine "technical means and organizational measures to ensure the timely detection of a crack in the coolant circuit pipelines" (ibid).

When developing primary circuit equipment and pipelines, design measures must be implemented to ensure monitoring of (NP-001-15; NP-089-15):

- pressure, temperature, chemical composition, as well as the level and flow rate of the coolant;
- explosive gas concentration;
- the presence of foreign objects;
- the content of neutron-absorbing nuclides in the coolant and supplied fluids;
- the activity of the coolant and other process fluids;
- possible leaks;
- the movement of elements due to temperature changes, etc.

The design documentation must specify the monitored parameters, the monitoring methods, locations for installations of sensors and devices for sampling process fluids, and also the monitoring frequency. To monitor the state of the reactor vessel metal under the influence of temperature and neutron irradiation, containers with witness samples must be placed in the reactor, and the design documentation "must include a Program for Monitoring Radiation Embrittlement and Thermal Aging" (NP-089-15).

The development of in-service monitoring and diagnostics in the nuclear energy industry is moving towards the creation of complex automated systems consisting of a number of local systems that perform the corresponding functions related either to individual elements of a nuclear facility (the reactor vessel, the main circulation pump, etc.) or to individual physical processes (e.g., related to coolant leak).

One such system is the vibration-noise diagnostic system, which is installed in all WWER-type reactors currently under construction (see Figure 4.2). It is capable of detecting significant features in the broadband mechanical vibrations that accompany reactor operation and identifying these features as signs of deviations from the normal state. In particular, changes in the frequencies and amplitudes of vibrations make it possible to draw the conclusions about (Arkadov et al., 2004):

- decreasing the stiffness of the elements by which the reactor shaft is attached to the reactor pressure vessel;
- similar degrading the spring blocks and spacer grids of fuel assemblies;
- wearing the fasteners (keys) through which the reactor shaft comes into contact with the reactor pressure vessel in the lower unit of their mechanical connection.

Some other capabilities of in-service monitoring and diagnostic systems, as applied to NPPs, can be found in the author's book (Spirochkin, 2019).

The use of in-service monitoring and diagnostics in aerospace engineering corresponds to specific features of this industry – see, for example, Xu and Xu (2017); Sause and Jasiūnienė (2021); Ranasinghe (2022). In particular, when operating a fleet of similar airplanes, due to weight limitations and an operational profile with repetitions of flights alternating with stays on the ground, most monitoring and diagnostic operations are performed in ground conditions. Modern planes and helicopters are equipped with systems that monitor the performance of their key elements (engines, aerodynamic controls, landing gear, etc.). During flight, these systems alert the crew to any abnormalities or malfunctions, but such monitoring is primarily for situational awareness purposes and is not intended to provide data on the actual system state, in-depth diagnostics, or life cycle management. One area of ongoing research is exploring the potential of in-service monitoring to determine the structural health of aircraft engines by analyzing vibrations generated by the interaction of their components with the airflow at the inlet (Gelman et al., 2020).

In-service monitoring and diagnostic systems are installed in spacecraft, but due to strict weight and telemetry bandwidth limitations, the number of sensors and the volume of data collected and transmitted to the mission control center are also limited. In deep space missions, signal latency becomes a significant factor, complicating the human operator control of a spacecraft or planetary rover. Therefore, the main operating mode can be an autonomous regime, including analysis of diagnostic signals and, based on these, automated decision-making on the spot. However, such decisions are more of a tactical nature. Analysis of the spacecraft state and management of its life cycle, albeit very limited, remain the prerogative of the operator.

5.5.3 Structural Health Monitoring and Condition Monitoring

The development and implementation of monitoring systems in various fields of technology follow two closely related, yet specific engineering directions: *structural health monitoring* (SHM) and *condition monitoring* (CM). The purpose of SHM is to ensure the structural integrity of a technical system, and its objectives are the detection of damage and deterioration, as well as the assessment of the exhausted service life and remaining service life. A significant portion of the operations within SHM are carried out using *non-destructive examination* (or *non-destructive testing*) methods (Cartz, 1995; Klyuev, 2003; Hellier, 2020). These methods allow for the examination of the surface and internal layers of an object (structural element, part, or other sample) without damaging it. Some methods require non-destructive testing personnel to use sensitive equipment and therefore require appropriate qualifications. The testing results are typically interpreted by humans.

CM focuses on machines (reciprocating and rotating mechanisms, such as pumps, engines, and electric generators) and their moving components and aims to detect signs of their malfunctions in order to schedule maintenance and prevent sudden failure. The data collected and processed within CM are mainly vibration parameters (Brown and Jensen, 2007; Mills, 2010; Ghazali and Rahiman, 2021).

CM is governed by international standards, for example, ISO 13373-1:2002, ISO 13379-1:2025, ISO 17359:2018, and national regulatory documents. In Russia, such documents used in general mechanical engineering include national standards GOST 32106-2013, GOST R ISO 13373-3-2016 and so on.

The application of SHM to aerospace structures is described in a number of publications (see, e.g., Yuan (2016)), and is regulated by a number of industry standards, including SAE ARP6461A. However, it would be premature to claim that the relevant technologies have been widely implemented in this area (Sause and Jasiūnienė, 2021).

The fields of SHM and CM may overlap when the subject of direct study is dynamic processes accompanying the operation of a technical system or excited during impulse or vibration tests. In particular, the methodology for dynamic response analysis is common.

5.5.4 Monitoring and Diagnostic Tools

A typical automated monitoring and diagnostic system includes:

- a distributed network of sensors recording raw data on loads and (or) the structure's response to them;
- a processor unit that processes sensor signals to obtain arrays of data on loads and response in the required form, as well as derived diagnostic information;
- communication channels through which sensor signals are transmitted to the processor unit;
- a human-machine interface through which the operator gains access to processed data arrays and diagnostic information;
- a power supply network for all devices used.

Depending on the type of technical system being monitored and diagnosed, as well as its operating profile, sensors can be permanently installed in the structure or connected to it only for the duration of diagnostic procedures. The operating principles and design of various sensors used to record loading parameters and structural response can be found in numerous publications on this topic – see, for example, Al-Baradoni and Groche (2022); Saga (2023); Reverter (2024); Wei and Wei (2025). The current phase of development of monitoring tools is characterized by the introduction of:

- fiber optic sensors that record temperature, pressure, strains and accelerations at multiple points of a structure using physical contact of the fiber with the corresponding physical field (Matias et al., 2017; Pendão and Silva, 2022; Guemes et al., 2025);
- wireless sensors that operate without power supply and transmit data via over-the-air communication channels (He et al., 2022);
- "smart" materials and structures capable of generating signals characterizing their state under load (Sun, 2015; Savi, 2024).

Along with the approaches listed, non-contact methods for obtaining raw data are also used, based on coherent optical methods for recording the response of a structure, such as speckle interferometry or holographic vibration interferometry (Razumovsky, 2011).

The processing unit can either be permanently installed in the technical system or be connected to it only during periodic system monitoring procedures. The latter is typical, for example, for aircraft monitoring and diagnostic procedures on the ground. In the case of permanent installation, the processor unit is implemented in the form of one or more microprocessors.

Data processing is carried out using downloadable software, but can also be performed at the hardware level by appropriate specialized electronic devices. Processing functions include filtering of raw data, transformation to the required physical units and scales, transformation of time series into frequency domain (Fourier transform or Fast Fourier transform), wavelet transform, and various diagnostic operations.

The human-machine interface in the monitoring and diagnostics system can also be organized on a permanent or periodic basis using standard terminal devices, in particular displays, and appropriate software.

In Sect. 5.3.1, special tests were mentioned that complement the capabilities of structural health analysis based on the response to operational loads. Response parameters typical for normal operating conditions may not provide information about some hidden anomalies and faults in a technical system. Detecting these may require methods for excitation and analysis of forced vibrations across the entire dynamic response spectrum. One such method, developed by the author and his colleagues, is active vibration diagnostics, described in Sect. 6.3. Excitation of test vibrations of a structure is possible using appropriate force exciters, which should be included in the automated monitoring and diagnostics system in addition to the elements listed at the beginning of this section.

As follows from the above, performing in-service monitoring and diagnostics during the operation stage places specific demands on the design of a technical system, including the installation of additional elements, as well as the organization of communication channels between them, and their power supply. On the other hand, the construction and operating conditions of the system under design determine the architecture and required characteristics of the automated monitoring and diagnostics system. Therefore, an important part of the design work is the integration of these two systems.

5.5.5 Advantages and Limitations

In-service monitoring and diagnostics can reduce aleatory uncertainty for each specimen of a serially produced system and increase the level of safety due to knowledge of the current state of the structure and actual loading history. This approach also provides for more accurate forecasting of its future based on the assessment of the remaining service life (under expected operating conditions). In addition, in-service monitoring and diagnostics facilitates to reduce epistemic uncertainty by detecting:

- hidden imperfections of the system at the operation stage (part of source 2 in Table 3.1);
- emerging properties (source 3) caused by possible negative changes in the state (operability) of the system;
- uncontrolled impacts and changes in operating conditions (source 4);
- possible manifestations of previously unknown factors (source 7).

However, this approach has some inherent problems. Firstly, the scope of monitoring and diagnostics at the current stage of technological development is limited to observing a relatively small number of safety-critical elements and recording a narrow set of key parameters. Otherwise, the complexity and energy consumption of the distributed structure of the automated monitoring and diagnostic system increase significantly, and its operation in real time becomes difficult. Another problem is the resistance of sensors, communications and processors to adverse operational factors, which must be maintained throughout the entire service life. This is particularly relevant for NPPs, as shutdowns for repairs or replacements of elements are associated with significant economic losses, and even more so for spacecraft on long-range missions, when repair or replacement of elements may not be possible.

In any case, recording physical processes is inevitably accompanied by errors, for the assessment and reduction of which the signals coming from the sensors must be periodically calibrated. In addition, a future forecast based on data collected during the past interval of the service life may prove incorrect if sudden, uncontrolled impacts and changes in operating conditions occur in a subsequent period. Finally, all decisions based on monitoring and diagnostic data are made by the system operator. This means that human factors with its possible negative manifestations are not excluded from the system control loop.

References

Ailor, W.H. and Patera, R.P. (2007) Spacecraft re-entry strategies: meeting debris mitigation and ground safety requirements. *Proceedings of the Institution of Mechanical Engineers*. Part G: *Journal of Aerospace Engineering*. Volume 221. Special issue paper, pp. 947–953.

Al-Baradoni, N. and Groche, P. (2022) Sensor-integrated structures in mechanical engineering: challenges and opportunities for mechanical joining processes. *Production Engineering*, 16:423–434.

AP-25 (2009) Aviatsionnye Pravila. Chast 25. Normy lyotnoi godnosti samolyotov transportnoi kategorii (*Aviation rules. Part 25. Airworthiness standards for transport category airplanes*), 3rd edn. Interstate Aviation Committee, Moscow (in Russian).

Arkadov, G.V., Pavelko, V.I. and Usanov, A.I. (2004) Vibroshumovaya diagnostika WWER (*Vibration and noise diagnostics of WWER*). (Edited by A.A. Abagyan). Energoatomizdat, Moscow (in Russian).

ASME PTC 19.1-2005 (2006) Test uncertainty. An American National Standard. ASME, New York.

Brown, D.N. and Jensen, T. (2007) Machine-condition monitoring using vibration analysis. The use of spectrum comparison for bearing fault detection – a case study from Alma Paper Mill, Quebec, Canada. Application notes. Brüel & Kjær, 2007. https://www.bksv.com/media/doc/bo0253.pdf. Accessed October 26, 2025.

Cartz, L. (1995) *Nondestructive testing: radiography, ultrasonics, liquid penetrant, magnetic particle, eddy current*. ASM International, Materials Park, OH.

Chen, H.-.P. and Ni, Y.-Q. (2018) *Structural health monitoring of large civil engineering structures*. John Wiley & Sons, New York.

Chisholm, J.C. (1994) Integrated plant automation, control and monitoring systems: a new heart for an old station. *International Conference on Life Management of Power Plants 1994,* Edinburgh, UK, pp. 227–235.

Cocron, A. and Aronhime, L. (2022) Risk, uncertainty and innovation. *NATO Review*. Opinion, analysis and debate on security issues, 14 April 2022. https://www.nato.int/docu/review/articles/2022/04/14/risk-uncertainty-and-innovation/index.html. Accessed May 30, 2025.

Committee on Science and Technology (1986) Investigation of the Challenger accident. Report of the Committee on Science and Technology. House of Representatives. Ninety-ninth Congress, Second Session, House Report 99-1016, October 29, 1986. U.S. Government Printing Office, Washington, DC. https://www.govinfo.gov/content/pkg/GPO-CRPT-99hrpt1016/pdf/GPO-CRPT-99hrpt1016.pdf. Accessed October 09, 2025.

David, L. (2025) Mysterious smoldering wreckage in Australian Outback is likely part of a Chinese rocket. *Space*, October 21, 2025. https://www.space.com/space-exploration/launches-spacecraft/

mysterious-smoldering-wreckage-in-australian-outback-is-likely-part-of-a-chinese-rocket. Accessed October 21, 2025.

Dynamika (1992) Analiz regionalnoi bezopasnosti territorii Rossiiskoi Federatsii s uchetom riska padeniya aviatsionnykh i kosmicheskikh ob'ektov. Itogovyi otchet po dogovoru № 12–92 s Goskomitetom RF po chrezvychainym situatsiyam (Analysis on regional safety of the territory of the Russian Federation taking into account the risk of falling aviation and space objects. Final report on the agreement No. 12–92 with the State Committee of the Russian Federation for Emergency Situations). Research and engineering company *Dynamika*, Korolyov, Moscow region.

Feynman, R.P. (1986) Volume 2: Appendix F – Personal observations on reliability of Shuttle. In *Report of the Presidential Commission on the Space Shuttle Challenger accident*. https://engineering.purdue.edu/~aae519/columbialoss/feynman-rogersrpt-app-f.pdf. Accessed October 09, 2025.

Gallai, M.L. (1985) S chelovekom na bortu (*With a man on board*). Sovetskii pisatel, Moscow (in Russian).

Garcia, M.A. (2024) NASA completes analysis of recovered space object. NASA | International Space Station, April 15, 2024. https://www.nasa.gov/blogs/spacestation/2024/04/15/nasa-completes-analysis-of-recovered-space-object. Accessed October 13, 2025.

Gelman, A. (2008) Objections to Bayesian statistics. *Bayesian Analysis*, 3(3):445–450.

Gelman, L., Petrunin, I., Parrish, C. and Walters, M. (2020) Novel health monitoring technology for in-service diagnostics of intake separation in aircraft engines. *Structural Control and Health Monitoring*, 27(5):1–14.

Ghazali, M.H.M. and Rahiman, W. (2021) Vibration analysis for machine monitoring and diagnostics: a systematic review. *Shock and Vibration*, 2021(2):1–25.

GOST 32106-2013 (2013) Kontrol sostoyaniya i diagnostika mashin. Monitoring sostoyaniya oborudovaniya opasnykh proizvodstv. Vibratsiya tsentrobezhnykh nasosnykh i kompressornykh agregatov (Condition monitoring and diagnostics of machines. Hazardous equipment monitoring. Vibration generated by rotodynamic pump and compressor units). Interstate standard. Standartinform, Moscow (in Russian).

GOST R ISO 13373-3-2016 (2016) Kontrol sostoyaniya i diagnostika mashin. Vibratsionnyi control sostoyaniya mashin. Chast 3. Rukovodstvo po diagnostirovaniyu po parametram vibratsii (Condition monitoring and diagnostics of machines. Vibration monitoring of machine condition. Part 3. Vibration parameter diagnostic guide). National Standard of the Russian Federation. Standartinform, Moscow (in Russian).

Guemes, A., Mujica, L.E., del-Río-Velilla, D. and Fernandez-Lopez, A. (2025) Structural health monitoring by fiber optic sensors. *Photonics,* 12(6):604.

He, Z., Li, W., Salehi, H., Zhang, H., Zhou, H. and Jiao, P. (2022) Integrated structural health monitoring in bridge engineering. *Automation in Construction*, Vol. 136, 104168.

Hejll, A. (2007) Civil structural health monitoring. Strategies, methods and applications. Doctoral thesis. Division of Structural Engineering, Department of Civil and Mining Engineering, Luleå University of Technology, Luleå, Sweden.

Hellier, C.J. (2020) *Handbook of nondestructive evaluation*, 3rd edn. McGraw-Hill Education, Columbus, OH.

IADC (2025) IADC space debris mitigation guidelines. United Nations Office for Outer Space Affairs (UNOOSA). Inter-Agency Space Debris Coordination Committee. IADC-02-01 Rev. 4, January 2025. Issued by IADC Steering Group and Working Group 4. https://www.unoosa.org/res/oosadoc/data/documents/2025/aac_105c_12025crp/aac_105c_12025crp_9_0_html/AC105_C1_2025_CRP09E.pdf. Accessed October 13, 2025.

ICAO (2018) Doc 9859. *Safety management manual*, 4th edn. International Civil Aviation Organization, Montréal, Canada.

Isac, M.D., Cîmpean, C. and Manea, D.L. (2025) The current status of structural monitoring: a bibliometric literature review. *Buildings*, 15(5), 739.

ISO 13373-1:2002 (2002) Condition monitoring and diagnostics of machines – Vibration condition monitoring – Part 1: General procedures. International Organization for Standardization, Geneva, Switzerland.

ISO 13379-1:2025 (2025) Condition monitoring and diagnostics of machine systems – Data interpretation and diagnostics techniques – Part 1: General guidelines. International Organization for Standardization, Geneva, Switzerland.

ISO 17359:2018 (2018) Condition monitoring and diagnostics of machines – General guidelines. International Organization for Standardization, Geneva, Switzerland.

Jones, A. (2021) Critiques of Bayesian statistics. Technical blog. https://andrewcharlesjones.github.io/journal/bayesian-critiques.html. Accessed January 04, 2026.

He, D., Cui, Y., Ming, F. and Wu, W. (2023) Advancements in passive wireless sensors, materials, devices, and applications. *Sensors*, 23, 8200.

Klyuev, V.V. (ed.) (2003) Nerazrushayuschii control i diagnostika (*Non-destructive examination and diagnostics*). Handbook, 2nd edn., corrected and supplemented. Mashinostroenie, Moscow (in Russian).

Lewallen, M. (2025) Houses & people have already been hit by space debris–and planes could be next. *GlobalAir.com*, August 22, 2025. https://www.globalair.com/articles/houses-and-people-have-already-been-hit-by-space-debris-and-planes-could-be-next?id=11105. Accessed October 11, 2025.

Matias, I.R., Ikezawa, S., Corres, J. (eds.) (2017) *Fiber optic sensors. Current status and future possibilities*. Springer.

Mills, S.R.W. (2010) *Vibration monitoring & analysis handbook*. British Institute of Non-destructive Testing, Northampton, UK.

NASA (2025) Orbital debris quarterly news. Volume 29, Issue 3, September 2025, pp. 9–10. https://orbitaldebris.jsc.nasa.gov/quarterly-news/pdfs/ODQNv29i3.pdf. Accessed October 12, 2025.

NOAA (2023) National Weather Service. Cold weather hazards. Miami – South Florida. National Oceanic and Atmospheric Administration. https://www.weather.gov/mfl/about_cold#:~:text=With%20adequate%20low%20level%20moisture,pets%2C%20friends%2C%20and%20neighbors. Accessed October 10, 2025.

NP-001-15 (2015) Federalnye normy i pravila v oblasti ispolzovaniya atomnoi energii. Obshchie polozheniya obespecheniya bezopasnosti atomnykh stantsii (Federal regulations and rules in the field of atomic energy use. General provisions for ensuring safety of nuclear power plants). Federal Environmental, Technological and Nuclear Supervision Service of Russia, Moscow (in Russian).

NP-089-15 (2015) Federalnye normy i pravila v oblasti ispolzovaniya atomnoi energii. Pravila ustroistva i bezopasnoi ekspluatatsii oborudovaniya i truboprovodov atomnykh energeticheskikh ustanovok (Federal regulations and rules in the field of atomic energy use. Rules for the design and safe operation of equipment and pipelines of nuclear power facilities. Federal Environmental, Technological and Nuclear Supervision Service of Russia, Moscow (as amended by Orders of Rostekhnadzor dated 17.01.2017 No. 11, dated 19.11.2019 No. 442) (in Russian).

NRCKI (1998) Istoriya atomnogo proekta. Vypusk 13 (History of the atomic project. Issue 13). National Research Center *Kurchatov Institute*, Moscow. https://elib.biblioatom.ru/text/kiae-istoriya-atomnogo-proekta_v13_1998/p1 (in Russian). Accessed October 06, 2025.

Oberkampf, W.L., Trucano, T.G. and Hirsch, C. (2003) Verification, validation, and predictive capability in computational engineering and physics. Report SAND2003-3769. Sandia National Laboratories, Albuquerque, NM, and Livermore, CA.

OSI (2023) Montreal recommendations on aviation safety and uncontrolled space object reentries. Outer Space Institute. The University of British Columbia, Vancouver, Canada. https://

outerspaceinstitute.ca/osisite/wp-content/uploads/Montreal-Recommendations-on-Aviation-Safety-and-Uncontrolled-Space-Object-Reentries.pdf. Accessed October 12, 2025.

Patera, R.P. (2008) Risk to commercial aircraft from reentering space debris. *AIAA Atmospheric Flight Mechanics Conference and Exhibit*, 18-21 August 2008, Honolulu, Hawaii. AIAA-2008-6891. https://arc.aiaa.org/doi/10.2514/6.2008-6891. Accessed October 13, 2025.

Patrushev, V.S., Pletnyov, I.V. and Spirochkin, Y.K. (1989) Aviatsionno-kosmicheskii trenazhyor (Air-spacecraft simulator). Patent SU 1799479, November 24, 1989 (in Russian).

Pendão, C. and Silva, I. (2022) Optical fiber sensors and sensing networks: overview of the main principles and applications. *Sensors,* 22(19):7554.

Persi, D. and Freedman, D. (1986) On the consistency of Bayes estimates. *The Annals of Statistics,* 14(1):1–26.

Pich, M.T., Loch, C.H., De Meyer, A. (2002) On uncertainty, ambiguity, and complexity in project management. *Management Science*, 48(8):1008–1023.

Presidential Commission (1986) Report to the President by the Presidential Commission on the Space Shuttle Challenger accident. Washington, DC, 6 June 1986. https://ntrs.nasa.gov/api/citations/19860015255/downloads/19860015255.pdf. Accessed October 09, 2025.

Ranasinghe, K., Sabatini. R., Gardi, A., Bijjahalli, S., Kapoor, R., Fahey, T. and Thangavel, K. (2022) Advances in Integrated System Health Management for mission-essential and safety-critical aerospace applications. *Progress in Aerospace Sciences*, 128:100758.

Razumovsky, I.A. (2011) *Interference-optical methods of solid mechanics*. Springer-Verlag, Berlin, Heidelberg.

Reverter, F. (2024) A tutorial on mechanical sensors in the 70th anniversary of the piezoresistive effect. *Sensors*, 24(11):3690.

Roulette, J. (2021) SpaceX rocket debris lands on man's farm in Washington. *The Verge*, April 3, 2021. https://www.theverge.com/2021/4/2/22364582/spacex-rocket-debris-falls-farm-washington. Accessed October 13, 2025.

Ruilin, P., Lei, Z., Li, W. and Ying, F. (2005) Real-time monitoring and integrated automation control in steel rolling plant. *TENCON 2005 - 2005 IEEE Region 10 Conference*, Melbourne, VIC, Australia, 2005, pp. 1–4.

SAE ARP6461A (2021) *Guidelines for implementation of structural health monitoring on fixed wing aircraft. Rev. A. Aerospace Recommended Practice*. SAE International.

Saga, S. (2023) Physical sensors: mechanical sensors. In R. Narayan (ed.) *Encyclopedia of Sensors and Biosensors* (First Edition). Elsevier, pp. 62–75.

Sause, M.G.R., Jasiūnienė, E. (eds.) (2021) *Structural Health Monitoring damage detection systems for aerospace*. Springer (Series: Springer Aerospace Technology).

Savi, M.A. (2024) An overview of smart material systems and structures. In M. Bittencourt, J. Labaki (eds.) *Proceedings of the 8th International Symposium on Solid Mechanics. MECSOL 2022.* Lecture Notes in Mechanical Engineering. Springer, Cham.

Sivasuriyan, A., Vijayan, D.S., Górski, W., Wodzyński, Ł., Vaverková, M.D. and Koda, E. (2021) Practical implementation of structural health monitoring in multi-story buildings. *Buildings*, 11(6), 263.

Smith, C.A. (2002) *Automated continuous process control*. John Wiley & Sons, New York.

Spirochkin, Y.K. (1987) Metodika raschyota dinamicheskogo nagruzheniya gruzhov pri avariinoi posadke letatelnogo apparata (Method for analyzing dynamic loads on items of mass during an aircraft emergency landing). Dissertation. NPO *Energia*, Kaliningrad (now Korolyov), Moscow region (in Russian).

Spirochkin, Y.K. (1988) Iskhodnye dannye na provedenie eksperimentalnykh issledovanii parametrov reaktsii posadochnoi poverkhnosti s ispolzovaniem zhyostkikh maketov. Otchet P 26478-017 (Baseline data for experimental studies on the landing surface response parameters using rigid mockups. Report P 26478-017). NPO *Energia*, Kaliningrad (now Korolyov), Moscow region (in Russian).

Spirochkin, Y.K. (1993) Computer simulation of structural dynamics in accidents. *Mathematical Models and Computer Simulations*, 5(6):85–103 (in Russian).

Spirochkin, Y.K. (2019) Bezopasnost rossiiskikh AES s tochki zreniya inzhenera-mekhanika (*Safety of Russian NPPs from the viewpoint of mechanical engineer*). SUPER Izdatelstvo, Saint Petersburg (in Russian).

Spirochkin, Y. (2024) What artificial intelligence brings to aviation: new opportunities or new risks? ResearchGate, 18.02.2024. http://dx.doi.org/10.13140/RG.2.2.19525.63203. Accessed October 10, 2025.

Stepnov, M.N. (1985) Statisticheskie metody obrabotki rezultatov mekhanicheskikh ispytanii: Spravochnik (*Statistical methods for processing mechanical test results: Handbook*). Mashinostroenie, Moscow (in Russian).

Sun, B.-H. (2015) *Smart materials and structures*. Cape Peninsula University of Technology, Cape Town, South Africa, and Swiss Federal Institute of Technology, Zürich, Switzerland. https://www.researchgate.net/publication/312027997_Smart_Materials_and_Structures.

Taleb, N.N. (2020) *Statistical consequences of fat tails: real world preasymptotics, epistemology, and applications* (Technical Incerto Collection). STEM Academic Press.

Taylor, J. (1965) *Manual on aircraft loads*. Pergamon Press, Oxford; London.

Vardanega, P.J., Webb, G.T., Fidler, P.R.A., Huseynov, F., Kariyawasam, K.K.G.K.D. and Middleton, C.R. (2022) Bridge monitoring. In *Innovative Bridge Design Handbook. Construction, Rehabilitaion and Maintenance*. Second edition. Edited by A. Pipinato. Butterworth-Heinemann, Oxford, UK. Part IX, Chap. 32, pp. 893–932.

Wang, P. (2004) The limitation of Bayesianism. Research note. *Artificial Intelligence*, 158(1):97–106.

Wei, X. and Wei, Y. (2025) Automatic control technology of mechanical equipment based on microelectronic sensor signal control. *Frontiers in Mechanical Engineering*, 11:1557863.

Xu, J. and Xu, L. (eds.) (2017) *Integrated system health management. Perspectives on systems engineering techniques*. Academic Press.

Yuan, F.-G. (ed.) (2016) *Structural Health Monitoring (SHM) in aerospace structures*. Woodhead Publishing.

6 New Approaches to Reducing Uncertainty

Expanding on the consideration of known and practically used methods for reducing uncertainty in the design of technical systems, this chapter is devoted to innovative approaches, currently in the research and development stage. One of these is a methodology for analyzing design data to identify potential sources of epistemic uncertainty that may manifest themselves during the operation of the system being developed. Another approach is the concept of information support for the system life cycle, which forms the innovative core of life cycle management based on actual state. This concept expands the possibilities for monitoring and diagnostics during operation, using, among other techniques, analytical reconstruction. An active vibration diagnostics method is also considered, which creates an additional opportunity to determine the actual state of a structure by detecting hidden defects. The chapter concludes by presenting a new design concept called the Smart ProActive Resilient System (SPARS). It combines defense in depth against predictable hazards, as well as in-service monitoring and diagnostics of anomalies, with the idea of a biologically similar (bionic) response of the system to adverse events, including unexpected ones, providing the ability to recover from destructive impacts.

6.1 Detecting and Analyzing Markers of Potential Sources of Epistemic Uncertainty

6.1.1 Special Checks of Design Data

As noted in Sect. 3.1, the system being designed and the conditions of its future operation may contain hidden features that can serve as markers of potential sources of epistemic uncertainty. The approach presented below is aimed at detecting such markers and analyzing them in order to identify these sources and find opportunities to reduce uncertainty. The rationale for this approach is based on a retrospective examination of some situations in the past, when "unknown unknowns" hidden at the design stage unexpectedly emerged during operation. As a result of experimental and theoretical studies, they were explained and eventually entered the category of "known unknowns". The essence of the proposed approach is as follows: despite the fact that the existing knowledge does not allow us to assume which unknown factors (existing beyond our knowledge) may emerge in a new design, some features of this design could indicate the potential for something unknown to appear, and these features are subject to preventive analysis.

DOI: 10.1201/9781003670575-6

Markers of potential sources of epistemic uncertainty can be detected using special checks of the array of information characterizing the designed system. The first of them is the *check for emerging undesirable properties* of the system. The elastic follow-up discussed in Sect. 3.1 is one of these properties – the manifestation of creep in metals at high temperatures in areas of discontinuity of geometric, and (or) mechanical characteristics of the structure and, accordingly, its stiffness. The criterion for the significance of this effect, when it should be paid attention to, can be considered an abrupt increase in the stiffness (or thickness) of one of the connected parts of the structure compared to the other by at least two times. The fulfillment of such a criterion in a joint indicates adverse rheological behavior of this joint under cyclic loading. Before explaining the cause of this behavior, this structural feature served as a marker of the source of uncertainty in the cyclic strength of the joint. Currently, the quantitative assessment of the elastic follow-up is mandatory according to regulatory documents – see, for example, RCC-MR (2007) or ASME (2010).

It cannot be ruled out that an abrupt change in the characteristics of a system in the connections of its elements (or components) may also serve as a marker for the emergence of certain other undesirable properties (source 3 in Table 3.1). To detect and analyze such markers, appropriate research is necessary. This research could form part of experimental and theoretical work to substantiate design decisions. It should be noted that the study of the specific properties of systems and the emergence of these properties is the subject of a specialized scientific and technical discipline – systems engineering, which has a well-developed methodological base.

6.1.2 Cliff-Edge Effects

Apparently, the most dangerous emerging property is the so-called *cliff-edge effect*. It describes an abrupt change in the behavior of a system, including the time profile of the process related to that behavior, which occurs when some influencing parameter deviates slightly from a certain characteristic value. As a result, the current operating situation significantly deteriorates. Cliff-edge effects can lead to irreversible consequences and impair the controllability of the system. In nuclear energy, any cliff-edge effect must be excluded at the NPP design stage as a potential cause of deterioration in safety during operation (NP-001-15; IAEA, 2016).

Cliff-edge effects can be inherent not only to systems formed by a number of structural elements. Sometimes they appear in simpler cases, when a system is formed by a combination of a single structural element, its environment (boundary conditions) and applied load. Let us demonstrate cliff-edge effects using two such examples.

As a first example, we consider the loss of stability of an element – a thin-walled cylindrical shell – under an axial compressive force. The diagram in Figure 6.1 shows the relationship between the compressive force applied to the end of the shell and the axial displacement of this end.

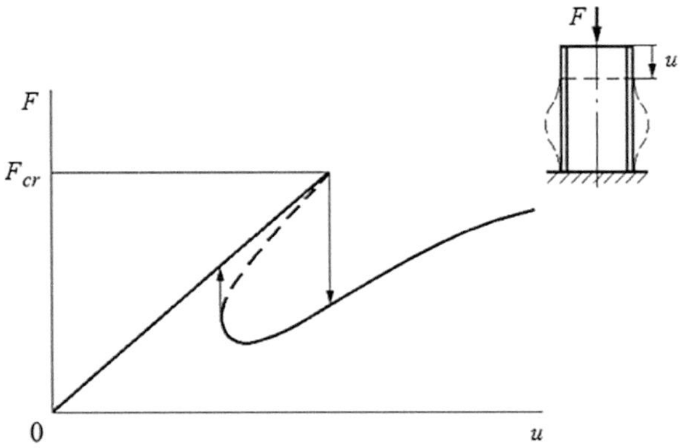

Fig. 6.1 An example of the cliff-edge effect: loss of stability of a cylindrical shell.

The following explanation can be given to this diagram. With an increase in *F*, at the beginning of this process, a stable compression of the shell occurs, accompanied by a proportional increase in the displacement, *u*. Upon reaching a critical value, F_{cr}, which characterizes the loss of stability, the proportionality between the applied force and the displacement is violated, the shell is instantly changes its shape, forming several buckling waves, and the deformation process "jumps" to the underlying branch of the *F*(*u*) curve. This jump is shown by the downward arrow. Within the framework of the mathematical theory describing such phenomena, the loss of stability in the form of bifurcation (and, accordingly, the abrupt loss of the load-bearing capacity of a structural element) is considered a catastrophe (Arnold, 1992). If there were no plastic strains during the compression and loss of stability, then the subsequent decrease in the compressive force *F* and accordingly displacement *u* first corresponds to the lower branch of the curve, and then is accompanied by a reverse jump to the overlying branch (upward arrow), at which the buckling waves disappear. The occurrence of plastic strains makes the loss of the bearing capacity of the structural element irreversible.

Instability with catastrophic jumps from one state to another is also characteristic of oscillations in nonlinear systems. They are demonstrated by the second example: resonance in a nonlinear elementary oscillator – the simplest mechanical oscillatory system with one degree of freedom, *x*. Its parameters are as follows: *m* (mass), *k*(*x*) (nonlinear stiffness), and *c* (damping coefficient). The oscillations are excited by a force *F* that changes over time (*t*) according to a sinusoidal law with a frequency *f* and amplitude equal to 1, as shown in Figure 6.2 (image a). With a slow increase in the excitation frequency, starting from very small values, forced oscillations with amplitude $A = x_{max}$ arise in the oscillator. The same figure (image b) shows the amplitude-frequency response curve of this oscillator, *A*(*f*), corresponding to the assumption of small damping, which does not have a significant effect on the value of the resonant frequency.

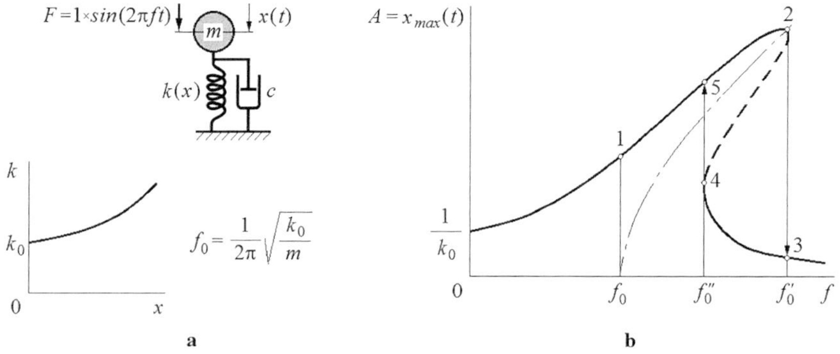

Fig. 6.2 Another example of the cliff-edge effect: resonance in a nonlinear oscillatory system: **a** nonlinear elementary oscillator, **b** the amplitude-frequency response curve.

If the system were linear, then as f approached the natural frequency of the system, f_0, resonance would occur, i.e. a sharp increase in A. However, the nonlinear system passes the corresponding point 1 without resonance. It occurs when $f = f_0'$ ($f_0' > f_0$), at point 2, and then the system jumps into the oscillation mode with a much smaller amplitude, corresponding to the lower, more stable branch of the amplitude-frequency response curve – to point 3. When moving in the opposite direction, i.e. slowly decreasing the frequency f and reaching the value f_0''($f_0 < f_0'' < f_0'$) at point 4, a jump into the oscillation mode with a larger amplitude occurs – to point 5.

When the exact properties of structural elements, including the values of the critical forces (loads), F_{cr}, the natural frequencies, f_0, the amplitude-frequency curves, $A(f)$, the jump frequencies f_0' and f_0'' are not known (and this situation is typical for the design stage), the conditions of loss of stability or resonances, as well as displacement values are difficult to predict accurately. However, this data can be crucial for ensuring strength, determining the parameters of control or diagnostic systems. This is why designers should try to identify cliff-edge effects of various natures in advance and prevent them using all available methods. Naturally, these effects may be related not only to structural mechanics but also to other physical processes within the system that determine its behavior, performance, controllability, and safety.

A marker of a potential cliff-edge effect – as one of the dangerous emerging properties or a source of possible unknown factors (the seventh source of uncertainty in Table 3.1) – can be considered a combination of the following aspects:

1) nonlinearity of the equations describing the behavior of the system under design (or its important element) in a certain operating mode;
2) the phenomenon of energy supply to the system (element) due to working processes or interaction with the environment;

3) detection in the state space of the system (element) of several possible states with close values of the main influencing parameter, but different levels of internal energy;
4) characterization of this state space either by an additional dimension free from applied loads, or by additional branches of state evolution; both options play the role of a channel through which excess internal energy can be released.

To the last aspect, it remains to be added that the release of the internal energy of the system through the above channel occurs very quickly; this process is practically uncontrollable and can have destructive consequences.

6.1.3 Consistency of the Design Basis

The next check of design information can be an analysis of the consistency of design basis: specifications, requirements of regulatory documents and reference data. The specific functions to be performed by the designed system, as well as its elements, and the specific values or ranges of values chosen as the object of monitoring and control under normal operating conditions must be clearly defined. They must not conflict with the requirements of regulatory documents, and must be substantiated by calculations, experiments, or reference data.

Regulatory documents are developed based on the generalization of many years of experience in the creation of technical systems and must objectively take into account modern technological advances. To this end, the existing provisions of these documents are periodically updated, but they are still characterized by inertia, meaning they may lag behind the current level of technological development. For example, ASME Boiler & Pressure Vessel Code, containing rules for design of NPP equipment and pipelines, is updated every three years. The "Consolidated List of Standardization Documents Used as Mandatory in the Russian Nuclear Energy Industry" is updated as necessary, but at least once every 5 years (Spirochkin, 2023). Thus, regulatory provisions regarding the use of new materials, unconventional design decisions, etc., may follow with a delay of 3–5 years after the initial introduction of innovations into advanced technical systems, including experimental ones. In some cases, the lag may be even more significant – for example, when it comes to ensuring safety in emergency situations. New data may not be sufficient to update existing overly conservative requirements, and the level of responsibility of specialists involved in rulemaking, and even more so of designers, who must adhere to the established rules, remains very high.

Moreover, extending existing requirements to new generations of technical systems may prove unjustified due to significant differences between the latter and the prototypes, whose features explicitly or implicitly formed the foundation for these requirements. Attempting such an extension may also reveal gaps in regulatory provisions through which uncertainty enters the design. Significant differences from prototypes and the resulting gaps can serve as markers of potential sources of epistemic uncertainty. A practical example of this kind is presented in Chap. 7.

Designers have several options to resolve such inconsistencies and, if not completely overcome epistemic uncertainty, at least absolve themselves of some responsibility. In the USA, one such option is to initiate research aimed at addressing the lack of regulation of a particular innovation within ASME. Such research typically results in the development and publication of a corresponding Code Case (ASME, 2010) – a document that legitimizes a successful solution to an emerging design problem until the next revision of the ASME Code is released. In Russia, the Federal Service for Environmental, Technological and Nuclear Supervision (Rostekhnadzor) permits design organizations working in the nuclear energy to substantiate design solutions not covered by regulatory provisions by submitting documentation containing methods and results of calculations or experimental data for a detailed expert review (Spirochkin, 2023). According to FAA rules, for aircraft of unconventional design or constructed using new materials, the level of crash protection must be substantiated by demonstrating equivalence with an existing prototype – "a traditionally configured metallic airplane" (Federal Register, 2015). The described opportunities for legalizing innovations that conflict with traditional design methods and regulatory provisions are based on the safety equivalence criterion and remove unnecessary barriers to technological development. The corresponding regulatory trend is discussed in more detail in Sect. 9.4.

Reference data can constitute a significant portion of the information used to support design decisions. It can include the design characteristics of prototypes, as well as their operating conditions (this information may sometimes be fragmentary). Therefore, another objective of the check described can be the consistency of reference data, with the aim of *detecting and analyzing unexplained phenomena and contradictory facts*. The same check should be made on the results of exploratory and other studies conducted to substantiate the design. The mentioned phenomena and facts, if discovered, can serve themselves as uncertainty markers – they indicate the presence of a source of uncertainty in the form of one or more unknown factors (source 7 in Table 3.1). The detection of such markers necessitates their analysis within the framework of ongoing experimental and theoretical work or the conduct of new research with the corresponding goals and methodology.

6.1.4 Complexity

Another procedure that can lead to the identification of potential sources of epistemic uncertainty is the *check for complexity*. It should be carried out not only with respect to the technical system being created, but also to two other systems associated with it: the production system and the operation system mentioned in Sect. 3.1. Complexity of any of them can cause unknown adverse factors to appear in life cycle processes. However, to act as a marker, complexity must be a measurable indicator, i.e. expressed quantitatively. A measure of system complexity can be considered a quantitative characteristic of the resources required for its monitoring, assessment of its state (operability), and control. Resources can include the intensity of relevant work operations, the resolution of sensors and processors, the parameters of control actions, etc.

If sufficient resources are available, the system does not create any particular difficulties for its support even in the case of a large number of elements included in it. If the amount of resources is inadequate for the tasks of monitoring, assessment and control, the system is not fully traceable or controllable and should be considered complex. Thus, complexity depends on the scientific and technical level, material and organizational support for the design and post-design stages of the life cycle.

Nonlinearity has already been considered above as one of the factors contributing to the emergence of cliff-edge effects that worsen the controllability of the system. Among the most adverse types of nonlinearity in this regard, the following should be noted:

1) nonlinearity with positive curvature in the control loop, when the gradient of the system response, *Ɽ*, to the control action, *Ç*, increases with increasing argument, as shown in Figure 6.3, image a;
2) relay-type nonlinearity present in the dependence of some important property of the system, *P*, on the spatial coordinate or time, *ȥ*, – see the same figure, image b.

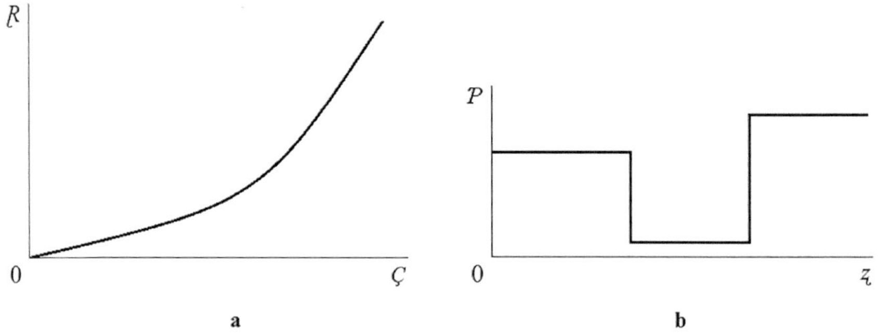

Fig. 6.3 Two adverse types of nonlinearity: **a** nonlinearity with positive curvature, **b** relay-type nonlinearity.

It is the detection of these types of nonlinearity that is the main goal of the *check for nonlinearity* due to their potential contribution to epistemic uncertainty. The corresponding markers are positive curvature and the above-mentioned peculiarities of functions describing the properties of the system that are important for its performance, controllability, or safety.

It is noteworthy that the second type of nonlinearity not only can create a cliff-edge effect but is itself similar in form to it. There may be a profound relationship between three adverse system characteristics – the potential for cliff-edge effects, complexity, and nonlinearity. The presence of such features in the system under design, i.e., the presence of the markers described above, points to possible sources of uncertainty listed in Table 3.1 under numbers 3 and 7.

6.2 Life Cycle Management Based on Actual State

6.2.1 The Motives for Developing a New Methodology and Its Main Components

When describing in-service monitoring and diagnostics in the previous chapter, the possibility of managing the life cycle of a technical system based on knowledge of its actual state was mentioned (Sect. 5.5.1). It is obvious that a priori assumptions about the adverse impacts on a system during operation and the rate of deterioration of its properties, specified in regulatory documents or taken from reference information (both sources are based on previous statistical data), are applicable to each specimen of a serially produced system, and especially to a unique system that has no analogues, with some error. The presence of such an error generates uncertainty, which may lead either to ineffectiveness of a new system due to excessive consumption of structural materials to ensure strength (because of excessive conservatism of regulatory provisions or inevitable caution when taking into account reference data) or to a safety deficit (in the case of rare extreme events not specified in regulatory documents and absent from references). The approach to system life cycle management described below represents an advanced and evolving methodology aimed at reducing this uncertainty and, consequently, improving safety and economic efficiency. It includes a set of organizational and technical measures covering all stages of the life cycle, from design to decommissioning.

The main functional components of life cycle management based on actual state are:

- in-service monitoring and diagnostics;
- analysis of the obtained data, formation of management decisions, their substantiation and adoption;
- implementation of the adopted decisions;
- information support for the life cycle.

The first three components are typical for conventional support of the operation of safety-critical and (or) expensive technical systems. The current state of development of in-service monitoring and diagnostic systems is described in Sect. 5.5. Modern automated monitoring and diagnostic tools can operate almost continuously or at specified intervals throughout the entire service life of a technical system.

The second of the listed components is primarily associated with:

- an assessment of the actual damage that has occurred in the structure and other subsystems subject to degradation during the analyzed period of operation (assessment of the exhausted service life) and

- a forecast of the further interval of safe operation (residual service life) assuming the predictability of all operational factors.

Within the framework of the conventional support, these activities are performed at characteristic moments of the operation stage: when approaching the end of the designated or previously extended service life, after extreme events, or after the completion of the current periodic monitoring interval. Analytical work is performed by the analysis engineers, typically from the design organization. They use in-service monitoring and diagnostic data, as well as additional information about system elements and operational processes obtained from the operating organization's personnel. Data from other stakeholders, such as equipment suppliers or service providers, can also be considered.

Based on the assessment of the exhausted service life and the forecast of the residual service life, as well as taking into account other aspects (economic, social, etc.), authorized representatives of the operating organization make decisions on the life cycle management of the technical system they operate. Possible decisions are selected from the following set: continued operation (including extension of the designated or previously extended service life), modernization, repair or replacement of elements that have failed or are nearing the end of their service life, cessation of system operation and decommissioning if it is impossible to repair or replace crucial end-of-life elements, and changing the operating profile. A change in operational profile can take the form of transferring a system whose key elements have reached the end of their service life to a different, more gentle mode of use, for example, as a facility for personnel training, research, etc.

Decisions regarding modernization, repair, replacement, decommissioning, or repurposing entail necessary communication, logistics, and production activities involving the operating organization and a number of other stakeholders. These activities are covered by the third functional component of life cycle management.

This is the methodology of the conventional support of the operation. Based on in-service monitoring and diagnostics, it replicates the advantages of this approach in relation to uncertainty reduction noted in Sect. 5.5.5. However, being limited to the first three components described above, this methodology also inherits the shortcomings of conventional monitoring. They manifest themselves in the following:

- the degree of reliability of assessments of the actual state (actual damage and exhausted service life) of a technical system and, consequently, the validity of decisions on life cycle management, depends on the completeness of monitoring and diagnostic data, which, in turn, is determined by the number of sensors and the spatial layout of their installation, and in this regard, every system has physical limitations;
- the assessment of the actual state is performed only for a few characteristic points in time, therefore changes in the state at intermediate intervals are not assessed, even if they are recorded by the monitoring and diagnostic tools and may pose a potential hazard.

In connection with the substantiation, adoption, and implementation of life cycle management decisions, another shortcoming of the conventional support methodology also becomes apparent: with a large volume of data characterizing the actual state of the technical system and life cycle processes, as well as with the exchange of this data between stakeholders, questions remain open regarding the completeness and consistency of the entire data set, the elimination of human errors in data processing, and the correct interpretation of data by each stakeholder.

Complex technical systems, such as NPPs, consist of many elements with different structures, working conditions, limit states to be considered, and loading histories. Therefore, the safety of the system during operation can be assessed using various indicators and methods. Some methods require extensive, specific, and detailed data, which exacerbates the shortcomings noted above. If the system was designed and constructed a long time ago, for example, before the 1990s, then its design data, as a rule, exists only in the form of paper documentation, and some of it may simply be lost, while the operational history is available in fragments due to an insufficiently developed monitoring system. From 2014 to 2016, the author led a team of Russian specialists involved in substantiating the extended operation of Units 5 and 6 of the Kozloduy NPP (Bulgaria) with WWER-1000 reactors. Each of the two units had been in operation for approximately 30 years by that time. Our team assessed their safety in the current state and for an extended period of up to 60 years using the leak before break concept, applying it to the primary circuit pipelines and adjacent pipelines, which are characterized by high stored energy (Spirochkin, 2019). The work was completed on time and to the required quality, but proved extremely labor-intensive due to the need to reconstruct missing data on these pipelines, which were important for the assessment, as well as to convert the data available in paper drawings or contained in other documents into a form suitable for analysis using modern computer tools.

The fourth component of the evolving life cycle management methodology, information support for the life cycle, aims to address the aforementioned shortcomings. This component of the methodology is innovative and has become its main distinguishing feature at this stage of development. It has practically become a separate, specific concept, which is presented below.

6.2.2 Information Support for the Life Cycle

The concept of information support for the life cycle originates from the US Department of Defense's Continuous Acquisition and Life-cycle Support (CALS) initiative, launched in 1985. The purpose of CALS was to reduce the time to market for defense systems, reduce the total cost of ownership, including logistics costs, and improve the quality of these systems throughout their entire life cycle (NCO, 2000). Over time, the idea of such support began to penetrate into other industries, keeping its focus on reducing the overall costs of creating and maintaining various complex technical systems by providing all stakeholders (involved in the life cycle) with a single source of system data. In particular, this source makes it possible to minimize errors in data transfer between stakeholders and from one stage of the life cycle to another (Kolchin et al., 2002; Norenkov and Kuzmik, 2002).

The main components of information support are:

- integrated information environment (IIE);
- information models of the system and its life cycle processes;
- data processing technologies;
- resource and process management technologies.

The first of these components is intended to store and transmit data about the technical system generated during its life cycle. IIE is a set of distributed databases, telecommunications channels, and terminal devices (computers of stakeholders). It ensures the accuracy, relevance, and security of data, as well as access to it by stakeholders who need it and are permitted to do so.

Data is stored in IIE in the form of information models. An information model is a data set, formed according to rules accepted by stakeholders, that characterizes an entity (material object, phenomenon, or process), as well as the relationships between entities. Such models are the primary source of data about a technical system and its life cycle processes. The category of information models can include a three-dimensional geometric model of a technical system taking into account its hierarchical structure (i.e., division into elements, assemblies (units) and parts), mathematical models used in various analyses, diagnostic models applicable to identifying deviations in the operation of a system, as well as other ordered sets of data characterizing life cycle processes.

A geometric model of a technical system, reflecting its structure, is created during the design stage and can be subject to change during the post-design life based on the results of quality control at manufacturing, in-service monitoring and diagnostics data, as well as information obtained during other life cycle operations. Based on the initial geometric model, a set of design mathematical models is constructed, including finite element models intended for analyzing the behavior of the system's structure under operational loads. During the post-design stages of the system's life cycle, the design mathematical models are validated using the results of various tests, as well as their parametric identification and updating are carried out in accordance with in-service monitoring data. To develop unified models available for data exchange and processing using different computing tools, a number of regulatory and reference documents are applicable, for example, the ISO 10303 (STEP) and ISO 15926 series of standards.

Data processing technologies, as well as resource and process management technologies are implemented using computer software tools that fall into the following categories:

- CAD (Computer Aided Design);
- CAE (Computer Aided Engineering);
- CAM/CAPP (Computer Aided Manufacturing/Computer Aided Process Planning);
- CNC (Computer Numerical Control);

- ILS (Integrated Logistic Support);
- LSA (Logistic Support Analysis);
- PDM (Product Data Management);
- PLM (Product Life cycle Management);
- ERP (Enterprise Resource Planning);
- WF (Work Flow), etc.

The concept of information support for the life cycle has apparently reached its greatest development at present in the nuclear energy industry. This was facilitated by the introduction at the same time of the Plant Life cycle Management (PLiM) strategy, driven by the need to ensure the long-term operation of existing and under-construction NPPs (IAEA, 2006; Kang et al., 2008; IAEA, 2021). An additional motivating factor is the importance and duration of the decommissioning stage of a NPP. During this stage, nuclear facility dismantling operations are carried out, the successful completion of which requires data on the actual structural health and condition of its elements and appropriate technological preparation.

In the Russian nuclear energy industry, information support for the life cycle was implemented in combination with the PLiM strategy. Initial work in this area focused on the design of nuclear transport reactors. The author participated in this work in 2006-2008 as the leader of the relevant R&D at NIKIET[1], as well as in its subsequent extension to equipment and pipelines of NPPs within the framework of the development of a number of regulatory documents (Spirochkin and Evropin, 2008; SPiR-O-2008; Evropin et al., 2009; Spirochkin and Atroshenkov, 2009; VERLIFE, 2013; OTT 1.5.2.01.999.0157-2013).

The main area of this R&D was the technological implementation of the principle of modeling a nuclear facility in its actual state and describing the real history of its operation. Methods were developed and tested for representing in a geometric model of a nuclear facility the initial technological imperfections introduced during the manufacturing stage of its elements and deviations from design documentation during installation of the facility, as well as defects arising during operation under impacts of adverse factors. Typically, initial technological imperfections and deviations from the design during installation are indicated in *as-built* drawings, but this method has not yet been implemented in the electronic geometric modeling technology.

Electronic geometric modeling of a nuclear facility in *as-built* state proved to be required when substantiating the extended operation of the Kozloduy NPP, mentioned in Sect. 6.2.1. Figure 6.4 shows, as an example, a three-dimensional geometric model of one of the analyzed pipelines, created during this work. The model corresponds to the actual geometry (routing) of the pipeline and the actual positions of all weld seams made during installation.

[1]NIKIET is the Russian abbreviation for 'Nauchno-Issledovatelskii i Konstruktorskii Institut EnergoTekhniki', one of the leading organizations in the Russian nuclear energy sector. Now, the name of this organization is JSC 'N.A. Dollezhal Research and Design Institute of Power Engineering'.

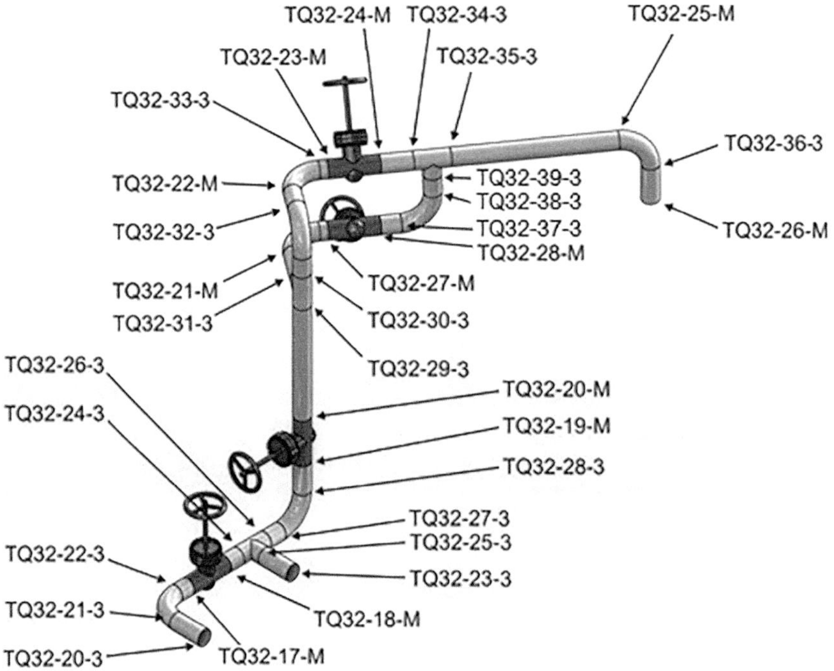

Fig. 6.4 3D model of the left line of the TQ32 pipeline at Units 5 and 6 of the Kozloduy NPP (alphanumeric indices designate weld seams).

As part of R&D on the implementation of information support for the life cycle, methods for constructing information models describing actual processes of the nuclear facility life cycle were also developed. One such model is the *actual operation model*, which shall accumulate data on the sequence of operating modes, their parameters (duration, temperature and pressure profiles in equipment and pipelines), as well as other internal and external impacts on the facility over the analyzed operating period. In addition, repairs and replacements of its elements shall be recorded in this model. This model is an alternative to the a priori *design operation model*. The technology of constructing the actual operation model, like the technology of as-built modeling, was tested during the substantiation of extended operation for the Kozloduy NPP.

Thus, information support for the life cycle is a concept that unites all its stages and all stakeholders through a comprehensive and shared set of data about the technical system, automated processing of this data, and continuous updating. This approach is not yet been fully implemented in any industry and is the subject of ongoing R&D. The incentive for this activity is the expected reduction in costs for the creation of complex high-tech systems in IEE and support of their operation. Project implementation timelines are also expected to be shortened. According to estimates made for the aerospace industry, certain indicators of financial and time costs can be reduced by up to 50% (Alemanni et al., 2008). These estimates can be questioned, but the advantage of information support for the life cycle is obvious compared to traditional paper-based collaboration, in which some data remains inaccessible to end users, other data is sometimes simply lost, and any changes require lengthy and complex approval procedures.

6.2.3 Opportunities Created by Information Support for the Life Cycle

Thus, in the framework of information support for the life cycle – the fourth component of life cycle management based on actual state – information models play a key role. Mathematical models, entering this category, which undergo parametric identification (parameter refinement) through pre-operational testing and are periodically updated based on in-service monitoring and diagnostics data, provide knowledge of the system's actual structural health and condition at each specific interval within the operation stage. This knowledge creates additional opportunities for reducing uncertainty compared to those provided by conventional in-service monitoring and diagnostics methodology.

The first of these opportunities is to overcome the shortcoming of this methodology noted in Sects. 5.5.5 and 6.2.1, which is associated with the relatively small number of directly observable and recordable parameters that characterize the state of the technical system and its operating conditions. Adequate approximation methods enable analytical reconstruction of load fields and response parameters based on fragmented data obtained from sensors (see Sects. 4.3.2 and 4.3.3), determining loading conditions for each element, and generating a real loading history. Updated mathematical models make it possible to analyze the actual state of load-bearing structural elements, including assessment of their exhausted and remaining service life. This facilitates more accurate safety assessments using various criteria.

The second opportunity involves addressing another shortcoming of the conventional support methodology (Sect. 6.2.1) – assessing the actual state and safety of the system only at a few characteristic points in time. With a proven technology for updating the mathematical model of the system and the actual operation model, such assessments can, in principle, be conducted at any other time and be tied only to the frequency of monitoring data updates, i.e., the frequency of recording sensor signals. This creates the possibility of early diagnosis (recognition) of adverse changes in the system's state and proactively changing the operating mode before serious deviations occur, not to mention the development of an emergency situation.

The third opportunity is also related to early deviation recognition and proactivity, i.e., it is close to the second one. It is about managing potentially hazardous situations caused by unexpected, uncontrolled impacts on the system and changes in operating conditions that go beyond those anticipated in the design operation model. Implementing this opportunity allows information support to be used to reduce the uncertainty generated by the fourth source indicated in Table 3.1.

It should be added that information support for the life cycle helps to reduce uncertainty associated with:

- the emergence of new properties (source 3 in our classification);
- inaccuracies in the substantiations of design decisions (source 6) and
- unknown factors (source 7).

6.2.4 Limitations of the Life Cycle Management Methodology

Despite the innovative nature and many advantages of life cycle management based on actual state, the methodology described above still has some limitations. These limitations stem from possible negative manifestations of human factors, embodied in the people involved in various management operations: personnel supporting the work of automated monitoring and diagnostic tools, analysis engineers assessing the actual state of the technical system and forecasting future operation interval, representatives of the operating organization making decisions, and other stakeholders. All these individuals can make unintentional errors, as well as exhibit inattention, carelessness, negligence, etc. Thus, due to the incomplete automation of management, in which humans also participate, the fifth source of uncertainty, indicated in Table 3.1, remains. In identifying this source, the author does not take a definitive position on the part of proponents of full automation, but only points out the possibility of the penetration of uncertainty of this origin into life cycle management processes.

The vulnerability of the concept of full automation lies in the fact that, at the current level of technological development, it is apparently impossible to completely eliminate errors in automatically performed operations during data processing within the framework of in-service monitoring and diagnostics, which are superimposed on errors in the recording of physical processes. Errors of both the first and second groups are associated with the sixth source of uncertainty. Thus, while some aspects of the described methodology can reduce uncertainty of this nature (see Sect. 6.2.3), others, on the contrary, contribute to it.

6.3 Active Vibration Diagnostics Method

6.3.1 Theoretical Basis

To detect defects that may arise during the manufacturing stage, as well as to monitor the integrity of equipment and pipelines of technical systems during operation, methods based on various physical processes are currently used: ultrasonic, radiographic, eddy current, acoustic emission, capillary, magnetic particle and others. None of these methods has an exhaustive degree of reliability (corresponding to the detection of all defects), and in the case of manual or partially automated execution procedures, they are subject to the influence of human factors (Klyuev, 2003). Moreover, for some elements that are not directly accessible during operation, all existing monitoring and diagnostics methods are applicable only at the manufacturing stage or during construction of the system, its modernization and repair, i.e., when the system is not operating. During operation, a relatively small number of critical elements are typically monitored, and a small set of parameters are recorded, characterizing their state and the performance of the entire system indirectly – through the structural response to loads (Sect. 5.5.1). The limitations of existing methods for monitoring and diagnostics during operation force designers to consider other approaches to obtaining the necessary experimental data.

For the detection of hidden defects in structures (as well as in mechanisms), the vibration diagnostics methods mentioned in Sect. 5.5.2 are very promising. In our understanding, structural defects are local deviations of the geometric characteristics and (or) mechanical properties of materials (Young's modulus, Poisson's ratio, density, internal friction parameters, etc.) of the system elements from their nominal values (presented explicitly in the design documentation or in the form of references to other documents). Deviations may also manifest themselves in changes in the boundary conditions corresponding to the connections of elements, including the movable joints in mechanisms. For defects, certain threshold parameters are usually specified by regulatory documents, which imply the possibility of their detection using known physical processes. Defects, caused by imperfections in manufacturing and assembly operations, poor conditions of transportation or storage of elements, as well as the influence of adverse operational factors, reduce the load-bearing capacity of the structure and (or) violate the normal functioning of the mechanism and, if uncontrolled, can have a negative impact on the safety of the system.

Hidden defects may manifest themselves as changes in the response of a structure or mechanism to applied loads compared to the values expected for the nominal state (corresponding to the design documentation). For a mechanical system with distributed parameters, the response has the form of a field, so the scale of such manifestations can be greater than the defects themselves. This makes it possible, in principle, to detect defects in cases where their direct observation is difficult – by recording the mechanical response in accessible areas of the structure or mechanism under loading conditions that create the required (for defect detection) response field. Dynamic loading conditions (under which structural vibrations are excited) are much more informative than static tests, since the parameters of the forced vibrations demonstrate changes in the elastic, inertial and dissipative properties of a structure or mechanism, i.e., those aspects that fall under the definition of defects.

This can be illustrated by the following relationships of structural dynamics, valid under the assumption of small steady-state vibrations and viscous structural damping. Forced vibrations of this kind, when using FEM for their analysis (see Sect. 4.2.2), are described by ordinary linear differential equations in the form of Eq. (4.22).

To solve them, we can use the Laplace transform:

$$\mathcal{L}\{u(t)\} = \{u(p)\} = \int_0^\infty e^{-pt}\{u(t)\}dt, \tag{6.1}$$

where p is a complex variable: $p = \alpha + j\omega$ (α is its real part; j is the imaginary unit; ω denotes, as before (Sect. 2.1.1), the circular frequency).

In this case, the nodal velocities and accelerations of the finite element ensemble are expressed through nodal displacements as follows:

$$\{\dot{u}\} = p\{u\}; \{\ddot{u}\} = p^2\{u\}. \tag{6.2}$$

Taking into account Eq. (6.2), the field of nodal displacements can be determined (up to the initial conditions) as:

$$\{u(p)\} = [K]^{-1}(\{F(p)\} - p^2[M] - p[C]), \tag{6.3}$$

where $\{F(p)\} = \mathcal{L}\{F(t)\}$.

A defect can be represented as a set of local deviations of the corresponding members of the inertia, damping and stiffness matrices of the finite element ensemble from the nominal values, $[M_0]$, $[C_0]$, $[K_0]$:

$$\begin{aligned} [M] &= [M_0] + [\delta M] = [M_0 + \delta M]; \\ [C] &= [C_0] + [\delta C] = [C_0 + \delta C]; \\ [K] &= [K_0] + [\delta K] = [K_0 + \delta K], \end{aligned} \tag{6.4}$$

where δ denotes the changes in the values of the matrix members due to the presence of the defect.

By introducing these changes into Eq. (6.3), we obtain an expression for the nodal displacements that takes into account the defect:

$$\{u(p)\} = [K_0 + \delta K]^{-1}(\{F(p)\} - p^2[M_0 + \delta M] - p[C_0 + \delta C]). \tag{6.5}$$

This expression can be given another form:

$$\{u\} = \{u_0\} + \{\delta u\}, \tag{6.6}$$

where the column vector $\{u_0\}$ contains the values of displacements corresponding to the nominal parameters of the structure (defect-free state), and $\{\delta u\}$ covers the deviations from these values caused by the defect.

As follows from Eq. (6.6), observation of the vibration displacement field of a structure including hidden defects, u, under dynamic loading makes it possible to detect them by comparison with the vibrational displacement field corresponding to the defect-free state of the structure, $\{u_0\}$.

6.3.2 Common Features and Limitations of Existing Vibration Diagnostics Methods

Vibration diagnostics is currently widely used for reciprocating and rotating mechanisms: pumps, water, steam and gas turbines, electric motors, generators, etc. To identify hidden defects, the frequencies and amplitudes of the kinematic parameters of vibrations occurring during their work are used. These kinematic parameters (vibration displacements, vibration velocities, and vibration accelerations) are measured at certain points of moving or non-moving parts and are assessed according to established criteria. The measurement and assessment procedures are governed by the relevant regulatory documents – see, for example, the standards cited in Sect. 5.5.3.

For nuclear reactor elements, vibration-noise diagnostics (Sect. 5.5.2) is used, based on a joint analysis of signals from accelerometers and neutron noise sensors (Arkadov et al., 2004). The most common vibration diagnostic methods applied within the framework of SHM or CM (Sect. 5.5.3) have the following common features:

1) the source of information about potential defects is the behavior of the structure (load-bearing elements or parts of the mechanism) under operating loads;
2) the recorded vibration characteristics are its frequency and kinematic parameters at a limited number of points;
3) the sensors are in direct contact with the structural element being monitored (placed on it).

These features limit the capabilities of SHM or CM in their current form. Not all defects manifest themselves through vibrations that occur in operating loading modes. Not all defects can be detected and reliably identified solely by changes in frequency or kinematic parameters at the sensor installation points. Exposure of sensors to temperature, chemical, and radiation factors in areas of high intensity can significantly reduce their service life. In such cases, sensor durability must meet special requirements comparable to those of the basic equipment metal.

6.3.3 Description of the Active Vibration Diagnostics Method

The active vibration diagnostics method, originally proposed in 2007, described in article Spirochkin et al. (2010) and protected by a Russian patent (Spirochkin and Atroshenkov, 2010), overcomes the above limitations. This method uses the dynamic response to a harmonic (sinusoidal) vibration load (point excitation force), applied specifically to obtain information about potential defects in the diagnosed structure (a combination of load-bearing elements, a single piece of equipment, a pipeline span, etc.). The loading parameters are deliberately selected to ensure maximum defect detection. This explains why this method is defined as "active" vibration diagnostics, as opposed to diagnostics based on the analysis of the response to operational loads, which is considered "passive".

When a sinusoidal force with a frequency corresponding to a specific natural vibration mode is applied to the structure being diagnosed, stationary vibrations with a significant amplitude are excited, i.e., resonance occurs. The spatial distribution of vibration amplitudes is recorded, reflecting the vibration mode shape. Such recordings can be performed for a range of vibration modes by gradually increasing the excitation frequency and achieving successive local resonances. The natural frequency of the mode, f_i, and the corresponding vibration mode shape, the eigenvector $\{\phi_i\}$, potentially contains information about existing defects if their influence manifests itself in a given mode – as a change in frequency, as well as changes in the mode shape values at certain points of the structure relative to the nominal values characterizing its defect-free state: $\{f_{i0}\}$, $\{\phi_{i0}\}$.

The active vibration diagnostics procedure involves scanning the vibration frequency range from low to high frequencies, observing successive local resonances and recording their characteristics. These characteristics include the experimental value of each resonance frequency and the distribution of vibration displacement amplitudes over the surface of the structure, which enables the identification of the corresponding natural vibration mode. Recording the vibration displacement amplitudes is carried out over the entire observed surface of the diagnosed structure using holographic interferometry (Razumovsky, 2011). The relief of the vibration displacement field is displayed through interference fringes corresponding to different amplitude levels, which these fringes similar to isolines (contour lines). Changes in the vibration mode shape caused by a hidden defect manifest themselves as distortions of the interference fringes relative to similar data typical of a defect-free structure: local convergence of the fringes and disruption of their monotony, including the appearance of kinks. The presence of such distortions indicates the presence of a hidden defect in this area of the structure.

Due to observing the entire accessible surface of a structure using holographic interferometry, the described method has no limitations on the number of sensors or their installation locations. Its fundamental feature is the reliance on mathematical modeling, which is used to calculate the natural frequencies and vibration mode shapes of a defect-free structure as a basis for comparative evaluation of experimental data. The natural frequencies and vibration mode shapes are calculated by solving the generalized eigenvalue problem corresponding to Eq. (4.22) under the assumption of no damping:

$$[K]\,[\Phi] = [\omega^2]\,[M]\,[\Phi], \qquad ..(6.7)$$

where $[\omega^2]$ is diagonal matrix of eigenvalues – squares of the circular natural frequencies ω_i^2, ($\omega_i = 2\pi f_i$; $i = 1, 2, \ldots n$; n is the number of vibration modes under consideration);

$[\Phi] = [\{\phi_1\}, \{\phi_2\}, \ldots \{\phi_n\}]$ is the matrix of vibration mode shapes (eigenvectors).

Excitation of structural vibrations using a sinusoidal force, searching for local resonances and recording the vibration amplitudes at the corresponding frequencies are operations typical for dynamic frequency tests carried out for launch vehicles and spacecraft (Gladkii, 1975). In English-language literature, such tests are referred to as "modal testing" (Evins, 2009). Their purpose is the validation of mathematical models used for dynamic calculations, as well as to obtain input data for adjusting control systems. Thus, active vibration diagnostics differs from modal testing in its goal: detecting hidden defects. Another important difference is that vibration parameters are obtained over the entire surface of the structure, whereas modal testing typically uses local accelerometers.

6.3.4 Experimental Testing of the Method

The capabilities of the active vibration diagnostics method were tested and confirmed experimentally in laboratory conditions. The experiment was conducted on a 200 mm long pipe with an outside diameter 60 mm and a wall thickness 1.5 mm. A blind, narrow cut was milled on the inner surface of the pipe, oriented circumferentially and simulating a crack-like defect, as shown in Figure 6.5, (image a). Experimental distributions of vibration displacements were determined by exciting vibrations in a certain frequency range by applying a sinusoidal force, achieving a number of local resonances and recording an interference pattern (hologram) for each of them on a photographic plate. These distributions were compared with the vibration mode shapes obtained as a result of calculations using a finite element model of the pipe, consisting of three-dimensional (solid) finite elements, with mesh refinement in the defect area (Figure 6.5, image b).

The calculated values of the natural frequencies and vibration mode shapes were determined by solving the generalized eigenvalue problem represented by Eq. (6.7). The calculated frequency values corresponding to the resonances identified in the experiment differed from the experimental values by 6–10%. These differences can be explained by the insufficient accuracy of the finite element model in representing the mass and stiffness characteristics of the equipment used in the experiment. The errors could have been reduced by refining the model, but in this case, they did not attract the research team's particular attention, as the primary goal of the experiment was to test the possibility of detecting signs of a hidden defect in the distributions of vibration displacement amplitudes and, accordingly, in the vibration mode shapes. This possibility was confirmed.

The defect located inside the pipe manifested itself in the distortion of the interference fringe patterns observed from the outside and displaying the distribution of the amplitudes of radial vibration displacements in at least two vibration modes, the holograms of which are reproduced in Figure 6.6, images a and b. These images show a part of the outer surface of the pipe; the height at which the defect is located is marked with a dotted line. Image c in the same figure presents the calculated vibration mode shape for the first "shell" mode of the pipe. It can be seen that the pattern of the displacement contour lines corresponds to the distribution shown in image b, and the defect is observable as a distortion of the inner contour line in its lower part. The visibility of the hidden internal defect on the outer surface of the pipe is due to the three-dimensional strain state of the pipe

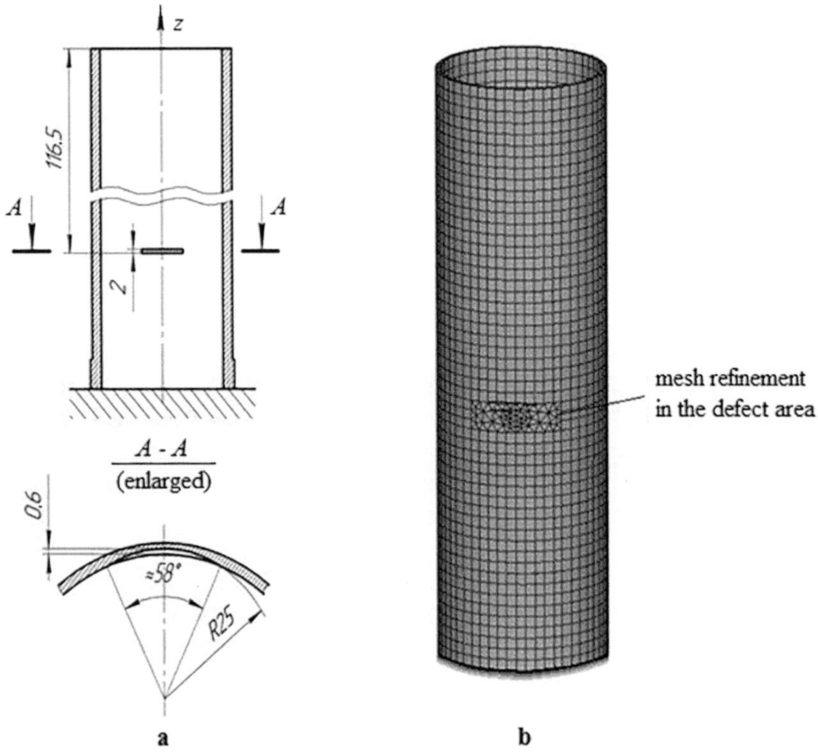

Fig. 6.5 Experimental testing of the active vibration diagnostics method: **a** the object of laboratory experiment and **b** its finite element model (rotated around the *z* axis by 180°).

wall, when the "opening" of the crack-like defect during bending vibrations leads to a change in the radial displacements recorded on the outer surface.

Some differences in the images b and c in Figure 6.6 are primarily due to different values of the displacement levels corresponding to the interference fringes (in experiment) and contour lines (in calculation). In addition, because the camera used to record the holograms was positioned approximately on half of the pipe high, its lower and upper parts were projected onto the photographic plate at a slight angle.

The comparison of experimental vibration displacement distributions with vibration mode shapes obtained through calculations is presented in Figure 6.7. It is performed by converting both quantities to the same normalized scale (from 0 to 1) and presenting their changes along the generatrix of the outer surface of the pipe, passing through the center of the defect. It is evident that:

1) the location of a hidden defect corresponds to a noticeable local disruption of the monotonic change in the amplitude of radial vibration displacements or the vibration mode shape observed on the outer surface for certain modes, and

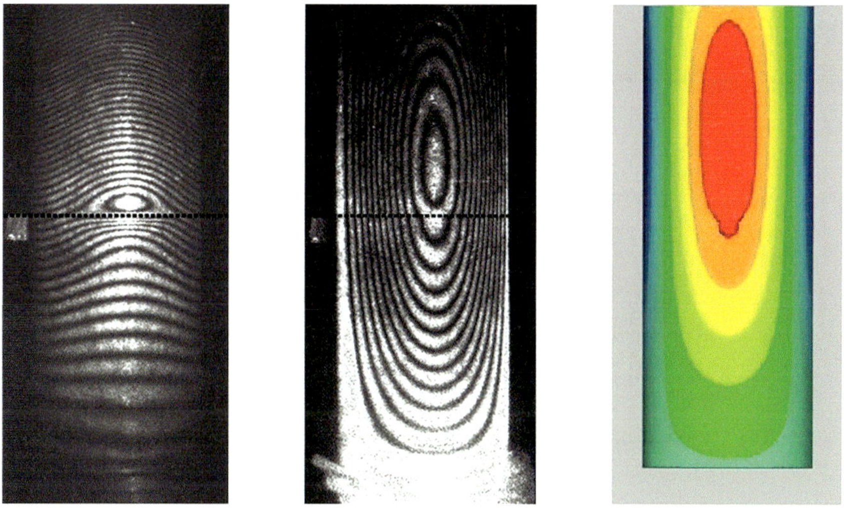

a b c

Fig. 6.6 Interference patterns of the radial vibration displacement amplitudes for: **a** the first "beam" mode ($f = 299$ Hz) and **b** the first "shell" mode ($f = 2370$ Hz); and **c** the calculated vibration mode shape for the first "shell" mode.

2) such local features of the corresponding experimental values and calculation results, which can be used to assess the presence of a defect, are in good agreement.

Thus, experimental verification in laboratory conditions confirms the feasibility and demonstrates the potential of the active vibration diagnostics method.

6.3.5 On the Applicability and Further Development of the Method

Active vibration diagnostics is undoubtedly applicable as a non-destructive examination method during the manufacturing of elements and construction of a technical system. Its advantage over other currently used methods of this type lies in the ability to record the structural response to the diagnostic input and detect defects without direct contact with the structure. Non-contact observation of vibration displacement field using holographic interferometry creates the basis for using this method directly during work of the system, when most common non-destructive examination methods cannot be applied at all. Thus, active vibration diagnostics increases the potential for reducing the uncertainty associated with hidden imperfections of the system structure (source 2 in Table 3.1).

The applicability of this method in the form described above may be limited due to the inaccessibility of some structural elements during system operation for exposure to coherent (laser) radiation and, consequently, the ability to obtain an interference pattern of vibration displacements. However, this limitation is not

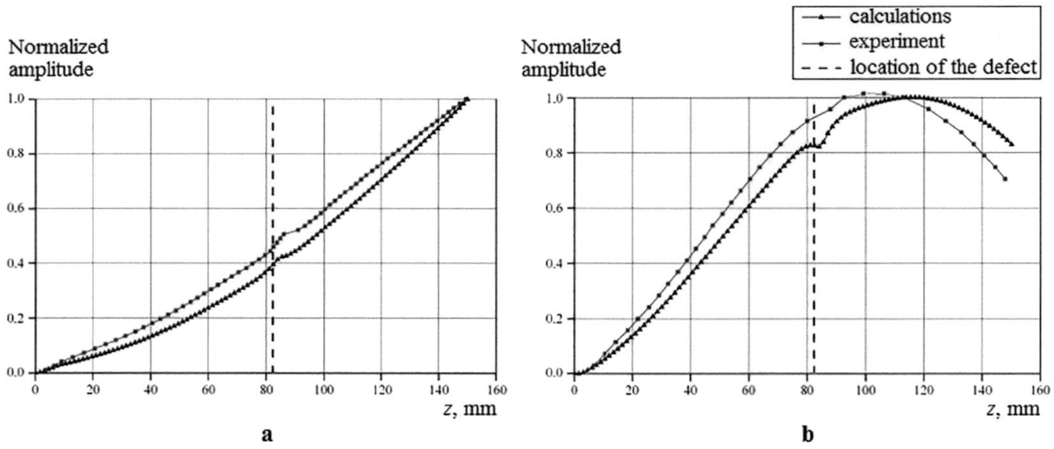

Fig. 6.7 Comparison of experimental and calculated distributions of radial vibration displacement amplitudes along the generatrix of the outer surface of the pipe for **a** the first "beam" mode and **b** the first "shell" mode.

due to the essence of the method, but to the technical means used to record the vibrational displacement field. In cases where laser interferometry is inapplicable, a solution to the recording problem should be sought among other methods that can provide data on the spatial distribution of vibration amplitudes under diagnostic loading. One possible approach is to integrate distributed sensor devices into the structure of safety-critical elements during their manufacturing. This approach relates to the developing area of "sensing" or "smart" materials and structures, which was mentioned in Sect. 5.5.4 with references to relevant publications.

Further improvement of the active vibration diagnostics method should be aimed not only at more accurately determining the location of anomalies hidden within a structure, but also at recognizing their type, shape, and size. Some prospects for such development are discussed in Sect. 8.2.

6.4 Introducing a New Design Concept: Smart ProActive Resilient System (SPARS)

6.4.1 The Main Content of the Concept

As is evident from the previous sections of this book, none of the currently used engineering approaches to reducing uncertainty in design completely solves this problem. Such incompleteness is particularly characteristic in relation to epistemic uncertainty associated with ignorance. The SPARS design concept (Smart, Pro-Active, Resilient System) presented below is based on the obvious point of view that it is impossible to eliminate uncertainty when creating technical systems intended for a long-term service under insufficiently known operating conditions that may conceal unpredictable events. Therefore, solutions to uncertainty-induced safety issues, as well as operational efficiency issues, especially when systems are complex, safety-critical, and expensive, should be sought by endowing them with properties similar to those of biological organisms – properties that increase the chances of survival in a dangerous, difficult-to-predict environment. These properties include: *smartness*, i.e. a sense (or an understanding) of the environment and internal state that enables the recognition of emerging threats, *proactivity*, meaning preparation for such threats in advance, and *resilience* – an idea that covers situational response to events, inter alia unexpected ones, together with the ability to reduce damage from adverse impacts and to recover.

The desired properties can be transformed into the following requirements: a system designed according to the SPARS concept must:

1) "observe" the environment, evaluate external conditions in comparison with acceptable ones (specified for design) and draw a conclusion about the "safety of being here";
2) "feel" its integrity, assess the health of elements and draw a conclusion about the "safety of being in the current state";
3) assess the possible danger of changes occurring outside and inside by comparing their quantitative characteristics with the system's resistance

potential based on actual (also subject to change) properties, and select a response mode that is acceptable from the point of view of "no loss of safety";

4) adequately respond to an adverse event (possible culmination of the above changes), i.e. implement actions aimed at minimizing the danger, for example, avoid extreme external impacts if the system is mobile and able to avoid them, or mitigate these impacts by absorbing and dissipating their energy;
5) assess in real time the development of the adverse event and the internal state according to safety criteria (assuming the possibility of an accident); if a safety deficit is detected, perform special actions to protect people and maintain the operability of elements important for safety, including localization and passivation of hazard factors, ensuring evacuation from the accident zone, as well as other actions that prevent the accident from escalating into a catastrophe;
6) assess the state of the system and the environment according to safety criteria after the adverse event; draw a conclusion on life cycle management applicable to the subsequent time interval, taking into account the efficiency criteria;
7) be able to recover, at least partially, if avoidance of extreme external impacts proved ineffective or their mitigation was such that one or the other caused damage to the system.

The design approaches implemented within the SPARS concept that meet these requirements are based both on ideas and technologies already used in safety-critical industries, as well as on a number of innovations. In the context of the existing scientific and technical base, these approaches correspond to the concept of multi-level protection (defense in depth) against hazard factors (Sect. 1.1.5). Further, going through the items of the above list, it should be noted that the first two requirements are close, on the one hand, to the widely used SHM and CM methodology (Sect. 5.5.3), and on the other hand, to the concept of information support for the life cycle (Sect. 6.2.2), which is in the development stage. The system properties that meet them, and the properties partially related to items 4, 5 and 6, are close to the principles of Integrated System Health Management (Xu and Xu, 2017) and Intelligent Health and Mission Management (Ranasinghe et al., 2022). The SPARS concept incorporates these principles and in turn develops them.

An important advance from the level of safety achievable by the traditional design approach (Sect. 2.1) to the desired higher levels can be considered the property of proactivity, which meets requirement 3. The idea of proactivity originates in socio-economic studies (Bateman and Crant, 1999) and has since begun to penetrate into engineering sciences (Bea, 2005), but it has not yet been sufficiently developed in the latter.

Achieving a higher level of safety is also facilitated by improving the reactive properties provided for in requirements 4, 5 and 6. These properties, given the high speed of physical processes characteristic of many accidents, can be implemented using AI. To reduce damage from extreme external impacts by absorbing and dissipating their energy, which is part of requirement 4, the principles of *crashworthiness* are incorporated into the SPARS concept. They were initially developed for military aircraft (Zimmerman and Merritt, 1989) and later were applied to various types of vehicles to ensure safety in accidents (Frings, 1998; Du Bois et al., 2000; Ambrósio, 2001; Spirochkin, 2023).

Requirement 7 is fulfilled based on the principles of *resilience engineering* (Hollnagel et al., 2006; Hollnagel et al., 2010). Their implementation within the SPARS concept includes supplementing the properties of crashworthiness, largely achieved through controlled destruction of selected structural elements, with another generally accepted technology – constructing a system from replaceable or repairable modules. Such modules should primarily serve as physical barriers that absorb the energy of extreme loads and localize the damage zone. They may also include other structural elements and safety devices that could be destroyed during an accident.

6.4.2 Background and Current Development

Building on the above achievements, the SPARS concept demonstrates advantages over the traditional design paradigm, as well as over more sophisticated approaches to addressing problems associated with adverse events that may arise during operation and are characterized by uncertainty, including its epistemic component. The origins of the SPARS concept can be found in R&D conducted in the 1980s at the Scientific and Production Association (abbreviated in Russian as NPO) *Energia*[2] during the creation of the Soviet space transport system *Energia-Buran*, similar to the American *Space Shuttle* (Figure 6.8). One of areas of this activity was to ensure safety during a possible emergency landing of the *Buran* aerospace vehicle. At that time, some approaches inherent in the crashworthiness concept were implemented (Spirochkin, 2023). Information on this activity, essential for understanding the SPARS concept, is also presented below in Chap. 7.

The subsequent development in this area was facilitated by the participation of specialists from TsAGI[3], the leading Soviet and Russian aeronautics research center (Spirochkin, 2023). Methods for analyzing the behavior of a structure during an accident have also found application in the International Thermonuclear Experimental Reactor (ITER) project (ibid).

Another source of inspiration for the development of the SPARS concept was R&D carried out in the Russian nuclear energy industry from 2006 to 2009 in the field of information support for the life cycle of nuclear transport facilities. This topic is presented in Sect. 6.2.2.

---

[2]Currently, the joint stock company (JSC) S.P. Korolyov Rocket and Space Corporation *Energia*.

[3]TsAGI is the Russian abbreviation for 'Tsentralnyi Aero-Gidrodinamicheskii Institut'. Current full name of this organization is the State Scientific Center – Federal State Unitary Enterprise 'Prof. N.E. Zhukovsky Central Aerohydrodynamic Institute'.

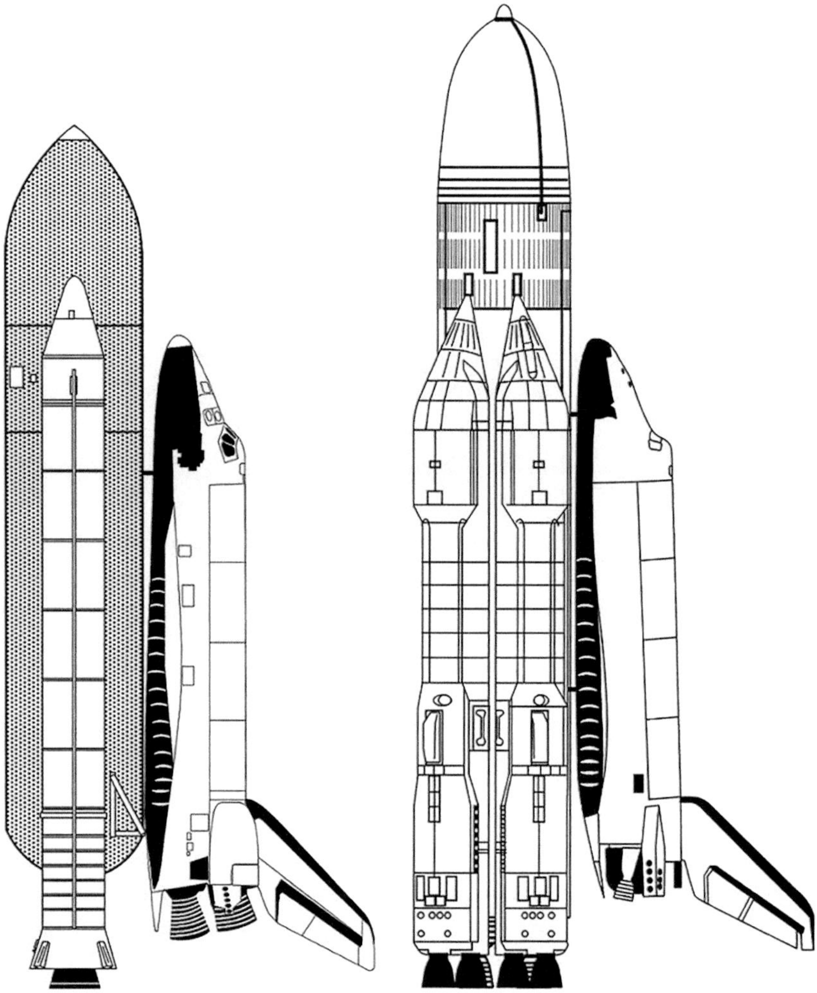

Fig. 6.8 *Space Shuttle* (left) and *Energia-Buran* (right). Reproduced from https://commons.wikimedia.org/wiki/Category:Illustrations_of_Buran_(spacecraft)#/media/File:Mir-72.jpg. NASA – Portree, David S.F. Original work of NASA – public domain.

The basic provisions of the SPARS concept were first outlined in the author's preprint (Spirochkin, 2024) and then in more detail in subsequent publications (Spirochkin, 2025a, 2025b). This concept appears to be one of the most effective tools for minimizing uncertainty in the design of complex, safety-critical, and expensive technical systems.

At this stage, the development of the SPARS concept is aimed at implementing the requirements formulated in Sect. 6.4.1 as they apply to aircraft design.

This development is primarily driven by alarming aviation accident statistics that indicate safety issues related to technological imperfections, the impact of increasingly complex, sometimes difficult to predict, flight conditions, and the negative manifestations of human factors. Currently, among Russian non-governmental organizations, the International Consultancy and Analysis Agency 'Aviation Safety'[4] analyzes aviation accidents and prepares recommendations on flight safety. Work to refine and test the SPARS concept is being conducted by the Society of Aerospace and Nuclear Engineers (SANE), a professional non-profit organization[5], with the direct participation of the author.

6.4.3 Innovative Aspects of the Concept

One of the innovative aspects of the SPARS concept can be considered the *interdisciplinary nature* of its approaches, allowing advances from one engineering field (industry) to be applied to another. This is precisely the case with the concept of defense in depth. It has been most developed in nuclear energy (see Sect. 1.1.5), and if applied to other industries, it could significantly improve safety. An illustrative example is provided in Chap. 7. The same is true for the principles of crashworthiness. Currently, they are used only in the design of mobile technical systems, i.e., vehicles. The use of the SPARS concept, which incorporates these principles, could expand their application area, e.g., to nuclear facilities.

Another innovation in the SPARS concept is bio-inspired design, in which bionic ideas are applied mostly to the behavior of the system, rather than to its structure, as is usually done. This aspect, which primarily covers the properties of smartness and proactivity (listed as required in Sect. 6.4.1), can be called *behavioral bionics*. Such properties are implemented in the designed technical system by equipping it with a special automatic control system that combines a sophisticated sensor network, advanced information processing devices, and actuators (devices that generate control forces) selected from a set that includes technologies being in use in other industries and, possibly, unfamiliar in this one.

The SPARS concept involves processing information using AI, implemented on the basis of neural network technology. The analysis of an aircraft emergency landing example presented in Chap. 7 demonstrates that a neural network approach is preferable not only due to the high speed of the accident development, to which the control system must respond, but also due to the ability to use heterogeneous information, the nature of which makes it difficult to apply classical mathematical methods. The use of neural networks constitutes the third innovative aspect of the SPARS concept.

Finally, the fourth innovation is design measures that facilitate the recoverability of a technical system after adverse events, but go beyond its modular construction. They increase resilience by improving physical barriers and safety devices

[4] https://aviasafety.ru.

[5] https://saneorg.wixstudio.com/index.

and include, for example, the use of *metamaterials* with advanced properties that traditional structural materials do not have. The metamaterials in question should provide sufficient energy absorption with minimal structural deformation, localization of hazard factors associated with uncontrolled movements of human-machine system elements in the event of an accident, as well as the ability of machine components to self-heal. The application of metamaterials for ensuring safety within the SPARS concept is an area of ongoing R&D.

References

Alemanni, M., Alessia, G., Tornincasa, S. and Vezzetti, E. (2008) Key performance indicators for PLM benefits evaluation: The Alcatel Alenia Space case study. *Computers in Industry*, 59(8):833–841.

Ambrósio, J.A.C. (ed.) (2001) *Crashworthiness. Energy management and occupant protection.* Springer: Wien; New York.

Arkadov, G.V., Pavelko, V.I. and Usanov, A.I. (2004) Vibroshumovaya diagnostika WWER (*Vibration and noise diagnostics of WWER*). (Edited by A.A. Abagyan). Energoatomizdat, Moscow (in Russian).

Arnold, V.I. (1992) *Catastrophe theory*. 3rd edn. Berlin, Springer.

ASME (2010) An International Code. 2010 ASME Boiler & Pressure Vessel Code. Section III. Rules for construction of nuclear facility components. Division 1, Subsection NH – Class 1 Components in elevated temperature service. American Society of Mechanical Engineers (ASME), New York.

Bateman, T.S. and Crant, J.M. (1999) Proactive behavior: meaning, impact, recommendations. *Business Horizons*, 42(3):63–70.

Bea, R.G. (2005) Design for reliability: human and organizational factors. In S.K. Chakrabarti (ed.) *Handbook of offshore engineering*. Elsevier, pp. 939–999.

Du Bois, P., Chou, C.C., Fileta, B.B., Khalil, T.B., King, A.I., Mahmood, H.F., Mertz, H.J. and Wismans, J. (2000) *Vehicle crashworthiness and occupant protection*. American Iron and Steel Institute: Southfield, MI.

Evins, D.J. (2009) *Modal testing – theory, practice and application*. 2nd edn. Research Studies Press Ltd., Baldock, Hertfordshire, England.

Evropin, S.V., Obushev, A.E., Spirochkin, Y.K. and Petrenko, A.V. (2009) Guidelines and dataware for life cycle management for NPP pipeline supports. *Proceedings of the 20th International Conference on Structural Mechanics in Reactor Technology* (SMiRT 20), Division 6. Design and Construction Issues, Paper 1774. http://www.lib.ncsu.edu/resolver/1840.20/23809.

Federal Register (2015) Vol. 80, No. 107/Thursday, June 4, 2015/Notices, pp. 31946–31948. Department of Transportation. Federal Aviation Administration. Aviation Rulemaking Advisory Committee; Transport airplane and engine issues; New task. https://www.faa.gov/regulations_policies/rulemaking/committees/documents/media/TAEcdeT1-06042015.pdf. Accessed October 16, 2025.

Frings, G. (1998) FAA Crashworthiness Research Program. *International Aircraft Fire and Cabin Safety Research Conference.* Atlantic City, NJ, 17 November 1998.

Gladkii, V.F. (1975) Prochnost, vibratsiya i nadezhnost konstruktsii letatelnogo apparata (*Strength, vibration and reliability of spacecraft structure*). Nauka, Moscow (in Russian).

Hollnagel, E., Woods, D.D. and Leveson, N.C. (eds.) (2006) *Resilience engineering: concepts and precepts*. Ashgate Publishing Ltd, Farnham, UK.

Hollnagel, E., Pariès, J., Woods, D.D. and Wreathall, J. (eds.) (2010) *Resilience engineering in practice.* Ashgate Publishing Ltd, Farnham, UK.

IAEA (2006) Plant life management for long term operation of light water reactors. Principles and guidelines. International Atomic Energy Agency, Vienna (Technical reports series No. 448).

IAEA (2016) IAEA Safety Standards Series No. SSR-2/1 (Rev.1). Safety of nuclear power plants: Design. Specific safety requirements. International Atomic Energy Agency, Vienna (Proceedings series).

IAEA (2021) Nuclear power plant life management. *Proceedings of an International Conference,* Lyon, France, 23–26 October 2017. International Atomic Energy Agency, Vienna.

Kang, K.-S., Inagaki, T., Clark, C.R. (2008) Engineering support for plant life management – The IAEA contribution. IAEA-CN-155-001KS. https://www-pub.iaea.org/MTCD/publications/PDF/P1362_CD/htm/pdf/Keynote%20Speakers/001KS.pdf. Accessed November 03, 2025.

Klyuev, V.V. (ed.) (2003) Nerazrushayuschii control i diagnostika (*Non-destructive examination and diagnostics*). Handbook, 2nd edn., corrected and supplemented. Mashinostroenie, Moscow (in Russian).

Kolchin, A.F., Ovsyannikov, M.V., Strekalov, A.F., Sumarokov, S.V. (2002) Upravlenie zhiznennym tsiklom produktsii (*Product life cycle management*). Anakharsis, Moscow (in Russian).

NCO (2000) *NATO CALS Handbook*. Draft 3. NATO CALS Office.

Norenkov, I.P., Kuzmik, P.K. (2002) Informatsionnaya podderzhka naukoyomkikh izdelii (*Information support for high-tech products*). N.E. Bauman Moscow State Technical University Publishing House, Moscow (in Russian).

NP-001-15 (2015) Federalnye normy i pravila v oblasti ispolzovaniya atomnoi energii. Obshchie polozheniya obespecheniya bezopasnosti atomnykh stantsii (Federal regulations and rules in the field of atomic energy use. General provisions for ensuring safety of nuclear power plants). Federal Environmental, Technological and Nuclear Supervision Service of Russia, Moscow (in Russian).

OTT 1.5.2.01.999.0157-2013 (2013) Opornye konstruktsii elementov atomnykh stantsii s vodo-vodyanymi energeticheskini reaktorami. Obshchie Tekhnicheskie Trebovaniya (Support structures for elements of nuclear power plants with water-water power reactors. General technical requirements). JSC Concern Rosenergoatom, Moscow (Contributors: Spirochkin Y.K., Obushev A.E., Filimonov S.V., Sorokin A.N., Tarakanov P.V. et al.). https://meganorm.ru/Index2/1/4293736/4293736524.htm (in Russian).

Ranasinghe, K., Sabatini, R., Gardi, A., Bijjahalli, S., Kapoor, R., Fahey, T. and Thangavel, K. (2022) Advances in Integrated System Health Management for mission-essential and safety-critical aerospace applications. *Progress in Aerospace Sciences*, Vol. 128, 100758.

Razumovsky, I.A. (2011) *Interference-optical methods of solid mechanics*. Springer-Verlag, Berlin, Heidelberg.

RCC-MR (2007) *Design and construction rules for mechanical components of nuclear installations.* Section 1 – Nuclear installation components. AFSEN, Paris.

SPiR-O-2008 (2009) Svod Pravil i Rukovodstv po Opornym konstruktsiyam elementov AES s WWER (Code of rules and guides on supports for elements of nuclear power plants with WWER). Standard of organization. ENES, Moscow (Contributors: Evropin S.V., Filatov V.M., Golovlyov Y.N., Obushev A.E., Rodchenkov B.S., Spirochkin Y.K. et al.) (in Russian).

Spirochkin, Y.K. (2019) Bezopasnost rossiiskikh AES s tochki zreniya inzhenera-mekhanika (*Safety of Russian NPPs from the viewpoint of mechanical engineer*). SUPER Izdatelstvo, Saint Petersburg (in Russian).

Spirochkin, Y. (2023) *Human factors and design*. Springer, Singapore.

Spirochkin, Y. (2024) Towards minimizing uncertainty in the design of technical systems. ResearchGate, October 2024. http://dx.doi.org/10.13140/RG.2.2.15766.15681.

Spirochkin, Y. (2025a) Aircraft design for safety in emergency landing. *Journal of Applied Material Science & Engineering Research,* Vol. 9, Issue 2, pp. 01-21, May 2025. https://www.opastpublishers.com/open-access-articles/aircraft-design-for-safety-in-emergency-landing.pdf. Accessed: September 29, 2025.

Spirochkin, Y. (2025b) Advanced concept for designing technical systems: SPARS (Smart ProActive Resilient System). *Design Problems and Knowledge Management*, 3(3): 25–47. https://saneorg.wixstudio.com/index/dpkm. Accessed: September 29, 2025.

Spirochkin, Y.K. and Evropin, S.V. (2008) Podkhody k informatsionnomu soprovozhdeniyu prochnosti YaU v techenie zhiznennogo tsikla (Approaches to information-driven support of the strength of nuclear facilities during the life cycle). In E.O. Adamov (ed.) *Annual report of NIKIET - 2008: Collection of articles*, pp. 120-121. NIKIET, Moscow (in Russian).

Spirochkin, Y.K. and Atroshenkov, R.S. (2009) Primenenie IPI (CALS-tekhnologii) v sovremennoi yadernoi energetike (Application of IPI (CALS technologies) in modern nuclear power industry). *Computer technologies for analysis of engineering problems in mechanics: 1st International Scientific School for Youth*, IMASh RAN, 9-13 November 2009. *Collection of lectures*, pp. 36-42. A.A. Blagonravov Institute of Machine Science of the Russian Academy of Sciences (IMASh RAN), Moscow (in Russian).

Spirochkin, Y.K. and Atroshenkov, R.S. (2010) Spocob diagnostirovaniya skrytykh defektov konstruktsii oborudovaniya i truboprovodov (Method for diagnosing hidden defects in equipment and pipeline structures). Patent RU 2,437,072; June 09, 2010 (in Russian).

Spirochkin, Y., Odintsev, I. and Atroshenkov, R. (2010) On active diagnostics method for assessment of technical condition of nuclear facility components. *Proceedings of the ASME 2010 10th Biennial Conference on Engineering Systems Design and Analysis* (ESDA2010), Istanbul, Turkey, July 12-14, 2010. Volume 2, Paper No: ESDA2010-25089, pp. 329–334. https://doi.org/10.1115/ESDA2010-25089.

VERLIFE (2013) Guidelines for integrity and lifetime assessment of components and piping in WWER nuclear power plants. IAEA, Vienna. Appendix F – Component and piping supports (Contributors: Obushev A., Spirochkin Y.).

Xu, J. and Xu, L. (eds.) (2017) *Integrated system health management. Perspectives on systems engineering techniques*. Academic Press.

Zimmerman, R.E. and Merritt, N.A. (1989) Aircraft crash survival design guide. Final report: Volume I (USAAVSCOM TR 89-D-22A) – Volume V (USAAVSCOM TR 89-D-22E), Simula Inc., Phoenix, AZ.

7 Application of the SPARS Design Concept to Ensure Safety in an Aircraft Emergency Landing

This chapter describes in detail one of the new approaches to uncertainty reduction currently being developed – the SPARS design concept. The description focuses on applying this concept to improve safety during an aircraft emergency landing. Such a landing is an important design case considered for all types of aircraft. Its key feature complicating safety assurance is that regulatory requirements to the aircraft structure, aimed at preventing serious injury to occupants (passengers and crew members), are combined with uncertainty in the loads acting upon impact with the landing surface and the inability to eliminate collapse of structural elements by traditional methods. To illustrate the capabilities of the SPARS concept, a hypothetical example of emergency landing of the Soviet aerospace vehicle *Buran* (analogous to the American *Space Shuttle*) is used. Along with the achievements of that time, the current state of work in this area within the SPARS concept, is also presented. This state is characterized by the use of AI in the form of a neural network, as well as the application of the principles of behavioral bionics.

7.1 Introduction to the Problem of Safety in Emergency Landing

7.1.1 Preliminary overview of the problem

Emergency landings of civil airplanes are rare, and each event of this kind not only attracts the attention of experts, but also causes public outcry, especially when it results in human fatalities. Two such disasters occurred at the end of 2024. On December 25, an Embraer 190 of *Azerbaijan Airlines*, Flight 8243 from Baku to Grozny, crashed while landing on the ground near the city of Aktau in Kazakhstan. Of the 67 occupants, 38 died (ASN, 2025a). On December 29, a Boeing 737-8AS of *Jeju Air*, flight 7C 2216 from Bangkok, made an emergency landing with the landing gear retracted at Muan Airport in South Korea and was completely destroyed. There were 175 passengers and six crew members on board. Only two survived (ASN, 2025b).

The possibility of emergency landing poses a serious challenge to aircraft designers. Airworthiness standards require that the designed structure ensure the safety of occupants in the event of a "minor crash landing", i.e. an emergency landing which results in minor damage and injury (CFR Part 25; AP-25, 2009;

DOI: 10.1201/9781003670575-7

CS-25, 2018). However, the conditions of such a landing and the loads acting on the aircraft structure when it hits the landing surface in an abnormal manner are uncertain. The behavior of this structure, the elements of which collapse on the impact, is difficult to predict. Even if regulatory provisions and design approaches ensure safety in the particular case mentioned above, in many other possible scenarios the chances of survival are not guaranteed and there is a safety deficit. The accidents mentioned above only confirm this regrettable judgment.

The shortcomings of the existing system of design measures to ensure safety in the event of emergency landing are even more clearly manifested in projects of aerospace vehicles such as the American *Space Shuttle* or the Soviet *Buran* (Figure 7.1) They were designed to transport astronauts, or cosmonauts, and large payloads into orbit and return them to Earth. On their return to Earth, these vehicles flew in the atmosphere like airplanes, using aerodynamic surfaces (wings, rudders, ailerons, etc.). Such vehicles had the basic features of airplanes and were considered their most advanced type in technological terms[1]. Therefore, the design specifications for *Space Shuttle* or *Buran* corresponded to the requirements for transport category airplanes. Among the unclassified documents, the provisions closest to these specifications are contained in the Soviet airworthiness standards (ENLG, 1985).

Since then, aerospace vehicles were designed that also use aerodynamic surfaces during the launch phase into space. New shuttles are being developed for commercial orbital flights and suborbital transportation of people and cargo[2]. Over the past decades, the requirements for transport category airplanes in the countries that produce such airplanes have undergone some changes, but the essence of the regulatory approaches has remained the same. There is no doubt that when creating new generations of aerospace vehicles, their belonging to this category will be preserved and the continuity of specifications will be ensured. However, the range of potential payloads (cargo) necessitates an expansion of existing requirements – along with the safety of occupants, the integrity of particularly valuable or dangerous payloads (for example, containing toxic, flammable or radioactive substances) must also be ensured and environmental pollution must be prevented, bearing in mind the possibility of damage during an emergency landing.

Thus, even a preliminary overview reveals the safety issues in the event of emergency landing. They are related to the uncertainty of the loads acting on the aircraft structure during a landing impact, the difficult-to-predict behavior of the structure under conditions of partial destruction and the lack of safety guarantees across the entire range of possible emergency landing scenarios. These issues are inherited by new generations of airplanes and aerospace vehicles under design.

---

[1] Aerospace vehicles may have design features related to their flight and the performance of specific operations in space, as well as re-entry into the atmosphere. However, from the perspective of safety in emergency landing, these features are insignificant.

[2] In aerospace flight, the terms "payload" and "cargo" can be synonymous.

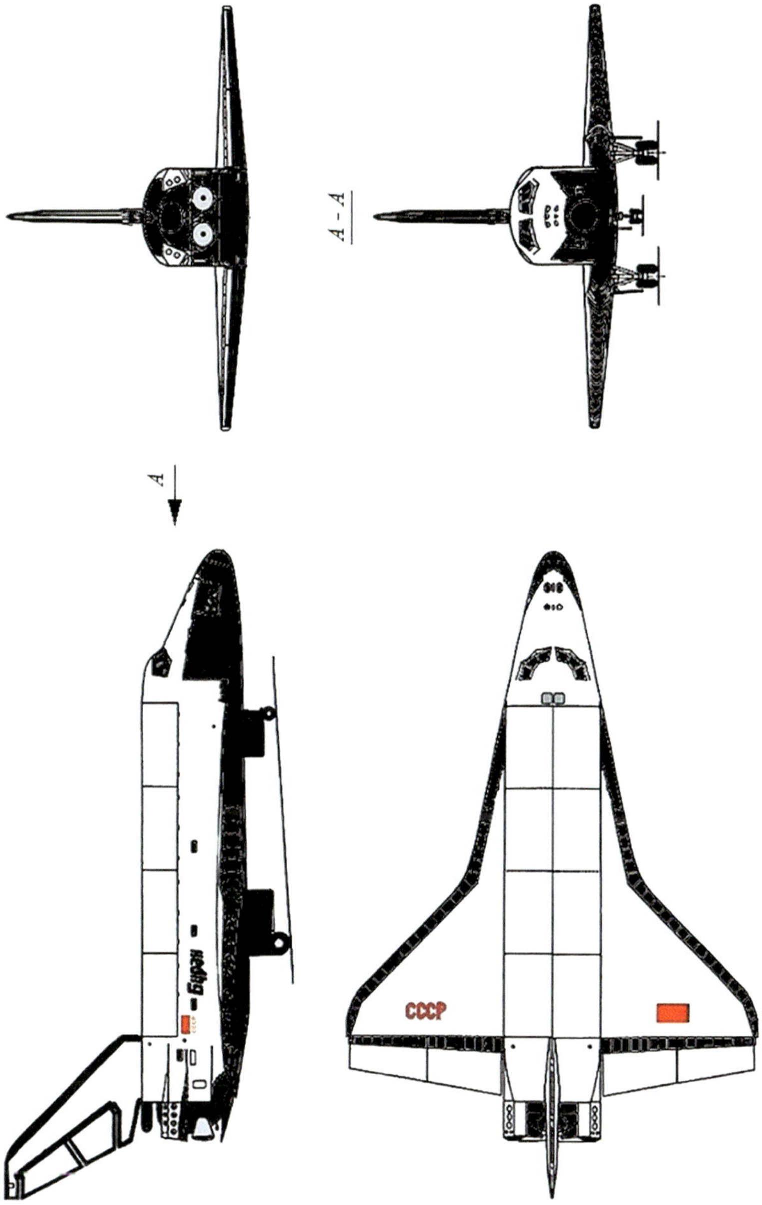

Fig. 7.1 Aerospace vehicle *Buran*. Reproduced from the image by Lee Brandon-Cremer: https://en.wikipedia.org/wiki/Buran_(spacecraft)#/media/File:04_OK-1.01.png, CC BY-SA 4.0.

For the latter, they are exacerbated by the risks associated with future payloads. Addressing these issues to improve safety is a pressing design problem.

Intensive R&D to improve safety in an aircraft emergency landing was carried out in the USSR during the creation of the Soviet space transport system *Energia-Buran* (Spirochkin, 1987). It was stopped in Russia with the closure of this program in the 1990s. The results achieved in this work, as well as the state of subsequent R&D in the world, are described in the book Spirochkin (2023). The corresponding activities have formed a separate engineering discipline – crashworthiness, mentioned in Sect. 6.4.1. However, advances in this discipline over more than three decades have focused primarily on military helicopters and ground vehicles, mainly automobiles (and to a lesser extent rail transport). They have not had a decisive impact on civil aviation or aerospace technology. There have been no significant changes in regulatory provisions or design practices to improve safety in emergency landing. This situation, close to stagnation, can be explained, in the author's opinion, by:

- insufficient awareness of the problem of safety deficit in emergency landings, given the relative rarity of such aviation accidents;
- a lack of breakthrough ideas for improving safety that go beyond crashworthiness principles (which are apparently already close to exhaustion);
- an insufficiently developed level of technology that could facilitate the emergence of fundamentally new ideas.

It is reasonable to expect that recent catastrophic events with civil airplanes, as well as the new prospects for the development of aerospace transport, will attract the attention of regulators and aircraft manufacturers to the design case "emergency landing". This chapter describes the continuation of the almost stagnant R&D on this topic and proposes to eliminate the safety deficit by applying the SPARS concept, introduced in Sect. 6.4. This concept has the potential to become one of the most effective tools for minimizing uncertainty and ensuring safety in the creation of complex, safety-critical, and expensive technical systems, including aircraft. It combines traditional design approaches that provide defense in depth against predictable adverse events with monitoring of external operating conditions and internal state of the system, new bionic ideas, and sophisticated information technologies, among them AI, to counter unexpected hazards.

7.1.2 Characteristic Features of Emergency Landing

As stated above, emergency landing is an important design case that must be considered for aircraft of all categories. Such a landing can be caused, for example, by birds entering the engines during climb, which leads to their shutdown (ASN, 2019), or by failure of the landing gear release system during descent (Sharma, 2024). When the engines stop working, there is little chance of returning to the departure airport or choosing another prepared landing strip. Therefore, pilots often make emergency landings on the nearest available land or water (ASN, 2009). Contact of the wheels of the extended landing gear with uneven, loose, or marshy soil may result in beyond-design loads acting on the mechanism components, and their destruction could cause serious damage to the aircraft structure (Joseph, 2012; Spirochkin, 2023). Although the best configuration for an airplane in emergency landing, including the position of the landing gear, is a subject of debate among professionals (Rossier, 2011), regulatory documents provide for the retraction of the landing gear – see, for example, CFR Part 25, § 25.561.

Thus, one of the characteristic features of the case "emergency landing" should be considered the direct contact of the fuselage with the landing surface. This feature justifies the use of the slang term "belly landing". It is important to note that emergency landing is *not a design case* for the fuselage structure[3]. The loads arising from contact may exceed the values allowed in design for normal operating conditions and cause destruction of structural elements. The combination of high inertial and contact loads in this situation determines the level of safety for aircraft occupants and the integrity of the cargo being transported, as well as the potential harmful impact of aircraft destruction on the environment.

Another characteristic feature of an emergency landing, as a specific operating mode, which is the subject of our analysis, is the preservation of aircraft controllability, at least partially. The entire process of an emergency landing with these features – from the moment the pilot makes the decision to perform it until the complete cessation of movement of the landed (or ditched) aircraft – can be divided into three phases:

1) pre-landing control;
2) main landing impact;
3) post-impact processes.

Table 7.1 provides values characterizing the duration of these phases, as well as the corresponding safety tasks.

Table 7.1 Emergency landing phases.

Phase	*Characteristic duration*	*Safety tasks*
1. Pre-landing control	Minutes	Ensuring a "soft" emergency landing
2. Main landing impact	From tens to hundreds of milliseconds	Limiting the intensity of hazard factors
3. Post-impact processes	From seconds to one minute	Counteracting dangerous post-impact processes

Control in the first phase of emergency landing, preceding the main impact of the aircraft on the landing surface, is carried out by the pilot, based on the state of the aircraft (primarily the operability of its engines and aerodynamic surfaces), environment data and recommendations from air traffic controllers. The

[3]The exception is structural elements that contribute to ensure buoyancy in ditching (emergency landing on water), such as external doors and windows.

expected result of control is to bring the aircraft to a chosen landing surface, the characteristics of which, in combination with the kinematic parameters (landing velocity vector and rotation angles of the aircraft axes), can ensure a "soft" emergency landing. "Softness" means avoiding fatalities, minimizing possible occupant injuries and aircraft damage. The first phase of emergency landing may also include touching down with less significant parts of the aircraft, for example, wings, to absorb excess kinetic energy by their destruction (Rossier, 2011). Thus, during emergency landing, several impacts are possible, but the main one is the impact of the fuselage, which is accompanied by the highest loads. Taking into account controllability in the first phase, emergency landing is sometimes called a "controlled crashing".

In the main landing impact phase (corresponding to intense dynamic contact of the fuselage with the landing surface and taking a fraction of a second), control by the pilot is impossible (Spirochkin, 2023). The hazard factors in this case are the following:

1) maximum values of accelerations acting on passengers and crew members, as well as acceleration profiles over time, since they can exceed human tolerance limits;
2) destruction of cargo and equipment attachment units under the action of inertial forces, which can cause uncontrolled movements of the torn off mass items with the threat of damage to the aircraft compartments and injury to people inside them;
3) injuries to people during their uncontrolled movements caused by contact with interior elements of the cockpit or passenger cabin;
4) deformations of the aircraft structure (relative displacements of its parts), which can lead to a reduction in the life volume of the cockpit and passenger cabin or contacts of large-sized cargo with structural elements of the cargo compartment, and also complicate the evacuation of occupants;
5) destruction of pipelines and tanks of the aircraft or transported cargo, which can pose a threat of fire, explosion and environmental pollution if they contain flammable, toxic or radioactive substances;
6) destruction of fuselage elements, leading to the aircraft sinking in the case of emergency landing on water, or ditching.

The listed factors are counteracted by a protection system, which includes a number of barriers (levels), which can be combined with passive or active safety means. The task of the protection system is to limit the intensity of hazard factors to values that exclude fatalities, reduce injuries to occupants and damage to the structure.

After the main landing impact, the aircraft continues to move along the landing surface with deceleration. This deceleration can be accompanied by such post-impact processes as continued deformation and destruction of structural elements, uncontrolled movements of occupants, cargo and equipment inside the compartments, and the release of contents of damaged pipelines and tanks.

The intensity of these processes and the degree of their danger largely depend on how the protection system worked during the main landing impact. Post-impact processes are counteracted by safety equipment that performs the functions of protecting, as well as mitigation and localization of consequences. If emergency landing is accompanied by significant negative consequences in the form of human injuries and destruction of the aircraft, it is characterized as a “crash landing”.

The duration of the main landing impact phase is the shortest, but it is not only associated with the most intense manifestations of hazard factors, but also determines the requirements for control in the preceding, first phase, as well as the course and outcome of the subsequent, third phase. Despite the short duration of the second phase, it is this phase that has a decisive influence on safety during an emergency landing. This phase should be considered the most important for ensuring safety, and it will be the focus of our attention.

7.1.3 Safety Issues

According to statistics, the frequency of emergency landings of civil airplanes does not exceed 10^{-5} 1/flight (Cronin, 2025). This frequency order corresponds to the boundary between *low probability events* (from 10^{-7} to 10^{-5}) and *moderate probable events* (from 10^{-5} to 10^{-3}) (AP-25, 2009). Such probability values, despite their smallness, cannot be ignored since fatal consequences are not rare (Rossier, 2011). Therefore, regulatory documents used by global manufacturers of airplanes and airplane-type aerospace vehicles provide for the possibility of emergency landing to be taken into account at the design stage (AP-25, 2009; CFR Part 25; CS-25, 2018). According to the relevant requirements of these documents, the designed aircraft must ensure the relative safety of each occupant. For example, in the American airworthiness standards this requirement is formulated as follows: “The structure must be designed to give each occupant every reasonable chance of escaping serious injury in a minor crash landing” (CFR Part 25, § 25.561). However, the conditions characterizing a relative safe, or “soft”, emergency landing (properties of the landing surface, components of the velocity vector and rotation angles of the aircraft axes), which affect safety in addition to the structure parameters, are not clearly defined in the regulatory documents. The choice of these conditions remains with the pilot. It should be noted that due to the high speed of the accident development and the psychological pressure exerted on the pilot by the perceived threat of a catastrophic outcome – see, for example, FAA-H-8083-3C (2021), Chapter 18, – not all aspects of the situation are within his control.

The diversity of potential emergency landing scenarios, the behavior of the aircraft structure beyond its strength limits, and the variability of human operator actions (including errors and inaction) can hardly be foreseen and taken into account at the design stage. Also, the question of the applicability of the specifications given in the regulatory documents in quantitative form and based, obviously, on statistical data, to aircraft of non-traditional design and manufactured using new materials remains open (Spirochkin, 2023). Thus, the approaches currently used to ensure safety in emergency landing cannot be considered reliable: at the design stage, safety is ensured largely in form, but not in substance (which is shown even more clearly below), and at the operation stage – mainly due to the qualifications of pilots, and only in the time interval preceding the main landing impact. This situation appears insufficiently manageable and does not correspond to the current state of science and technology.

7.2 Assessment of the Existing System of Protection Against Hazards in Emergency Landing

7.2.1 *Assessment Methodology*

The assessment of the existing system of protection against hazards in an aircraft emergency landing is carried out by determining how this system corresponds to the defense in depth concept described in Sect. 1.1.5. Such system should be formed by physical barriers, comprised of structural elements, in combination with safety means that are activated when these barriers are destroyed. It should also include measures implemented at all stages of the aircraft life cycle to maintain the effectiveness of the protection. The assessment involves analyzing how this system copes with the uncertainty inherent in a situation under consideration and ensures safety. The system of protection used at nuclear facilities, which is considered the most advanced, can serve as a benchmark for implementing the defense in depth concept in any other industry. It is worth recalling its characteristic features, expressing them in mathematical terms.

Nuclear facilities are equipped with multi-level protection against the main hazard – the release of ionizing radiation into the environment. The safety provided by such protection can be assessed using various indicators, and the number of independent physical barriers in the path of this hazard, N_{pb}, is one of them. At modern NPPs, this indicator reaches five (Sect. 1.1.5).

To assess safety on a probabilistic scale, each of these barriers should be assigned the probability of its failure $P_i (i = 1, 2, \dots N_{pb})$ or, conversely, the reliability value: $R_i = 1 - P_i$. Then we obtain the desired safety assessment in the form of the reliability estimate for the set of physical barriers:

$$R_{pb} = 1 - \prod_{i=1}^{N_{pb}} P_i. \tag{7.1}$$

Since each of the values of P_i is less than 1, the higher N_{pb}, the higher the value of R_{pb} and, accordingly, the safety (this provision has already been given in Sect. 1.1.5).

In addition to the use of physical barriers, the safety of nuclear facilities is ensured by other technical and organizational measures implemented at all stages of the life cycle, based on the possibility of deviations from normal operation and their escalation into accidents of varying severity. These measures form functional levels of protection, which number, N_{fl}, also is equal to five (Sect. 1.1.5). A probabilistic safety assessment taking into account functional levels of protection can be obtained using a mathematical expression similar to Eq. (7.1), but with input data characterizing the reliability of the implementation of each such level.

The number of functional levels of protection, N_{fl}, the number and physical parameters of safety means (active and passive safety systems), and quantitative data on inherent safety can be used, along with the values of N_{pb} and R_{pb}, as safety indicators for nuclear facilities. This entire set, combined with the advanced control, monitoring and diagnostic tools, as well as limitations on the manifestation of human factors, characterize nuclear energy as a relatively safe industry overall (Spirochkin, 2019). This characterization allows us to accept the defense in depth system cultivated in this industry as a basis for assessing the system of protection in the event of emergency landing of a transport category airplane (and, accordingly, an aerospace vehicle).

7.2.2 Depth and Reliability of Protection by Physical Barriers

In the event of emergency landing of a transport category (civil) airplane, there are two physical barriers in the path of the accelerations that constitute the first group of hazard factors (Sect. 7.1.2): the fuselage structure and the passenger or crew member seat equipped with restraining safety belts. The fuselage structure is not designed to withstand impact with a hard landing surface without the use of landing gear, so its behavior in emergency landing is a priori uncertain. However, the regulatory documents do contain provisions on the role of this structure as a barrier – see, for example, the citation from the American airworthiness standards (CFR Part 25), § 25.561 in Sect. 7.1.3. The cited requirement is associated with the correct use of seats, belts and other safety design provisions, retraction of wheels and limitation of accelerations acting on an occupant to the specified maximum values, $\{a_{max}\}$. Thus, in relation to the fuselage structure it is indirect. The requirements for a seat with restraining belts as a second physical barrier to resist acceleration are formulated in regulatory documents more clearly. In addition to specifying $\{a_{max}\}$, these requirements define the details of the acceleration-time profile, $\{a(t)\}$, for which seats and belts must be designed and tested, establish occupant injury criteria, and specify the impact resistance values and structural features – see, for example, CFR Part 25, § 25.562 and § 25.785.

Therefore, for the category of airplanes under consideration (which includes aerospace vehicles) the number of physical barriers that resist the first hazard factor can be estimated to be equal to two: $N_{pb}^{(1)} = 2$. However, there is uncertainty regarding the function of the fuselage structure as a barrier, and accordingly, its reliability in the event of impact with a hard landing surface is not determined. It should be noted that within the framework of the crashworthiness concept used in the design of military helicopters, a special energy-absorbing structure of the lower part of the fuselage is provided as the first physical barrier against impact accelerations (Jackson and Fasanella, 2003; Jackson et al., 2006; Lau et al., 2012; Guida et al., 2018). The behavior of this part is analyzed through calculations and experiments, which allows designers to minimize the mentioned uncertainty and obtain a sufficiently reliable first barrier.

The maximum acceleration values specified in the regulatory documents are based on experimental data from many years ago, presumably corresponding to the conditions of emergency landings of light airplanes. This hypothesis is supported by the ratios of ultimate acceleration components (CFR Part 25, § 25.561):

$$\begin{aligned}
&a_{x\,max} = 9\text{g when the forces of inertia act forward, and}\\
&a_{x\,max} = 1.5\text{g when they act rearward;}\\
&a_{y\,max} = 6\text{g when the forces of inertia act downward, and}\\
&a_{y\,max} = 3\text{g when they act upward;}\\
&a_{z\,max} = 3\text{g when the forces of inertia act sideward on the airframe, and}\\
&a_{z\,max} = 4\text{g when they act (sideward) on the seats and their attachments,}
\end{aligned} \tag{7.2}$$

where g is the acceleration of gravity; subscripts x, y and z denote the longitudinal, vertical and side components of the acceleration vector.

Emergency landings of light airplanes are characterized by short stopping distances and possible impacts with trees, bushes or other small obstacles (which contribute to a decrease in horizontal velocity), as well as frequent burying the nose in the soil (FAA-H-8083-3C, Chapter 18). These features determine higher values of longitudinal (directed along the aircraft axis) accelerations compared to vertical and sideward accelerations. The extension of these data to transport category airplanes (heavy-weight and wide-body), as well as to aerospace vehicles whose structure (in particular, the low curvature of the fuselage bottom – see Figure 7.1) and the kinematic parameters during landing differ significantly from those of light airplanes, is questionable. Thus, the maximum acceleration values specified in regulatory documents appear to be insufficiently substantiated when applied to modern aircraft, especially of new types.

The second group of hazard factors – the destruction of cargo and equipment attachment units and possible subsequent uncontrolled movements of the torn off mass items – is countered by physical barriers formed by the attachment units themselves, the structure of the cargo compartment and restraining structural elements. Regulatory documents set requirements for the strength of attachment units in the design case "emergency landing" so that the torn off mass items do not cause injuries to occupants, do not damage pipelines and fuel tanks, and do not impede evacuation. Strength requirements are expressed in terms of maximum values of inertial loads that the attachment units must withstand. They correspond to the maximum values of accelerations acting on occupants, $\{a_{max}\}$, taking into account the wear of the attachment unit parts – see, for example, CFR Part 25, § 25.561. The structure of the cargo compartment and restraining structural elements, including locking doors, must maintain the load-bearing capacity under relevant loads and also perform the function of localizing the possible consequences of the detachment of mass items (CFR Part 25, § 25.787 and § 25.789). Thus, the number of physical barriers in the path of the second group of hazard factors is three ($N_{pb}^{(2)} = 3$).

It should be noted that when expressing the requirements for the strength of the attachment units through the maximum values of accelerations, $\{a_{max}\}$, the effects caused by the dynamic properties of large-sized and heavy payloads are neglected. Their mass and stiffness of the structure can affect the dynamic response of the aircraft structure in emergency landing and, accordingly, the required strength of the attachment units (see Sect. 4.1.2). The provisions of the regulatory documents leave unclear the conditions under which such effects should be taken into account, and this uncertainty aggravates the above-mentioned shortcoming of specifying the required strength using $\{a_{max}\}$.

Injuries to occupants due to their uncontrolled movements during an emergency landing – the third group of hazard factors – are prevented by seats with restraining belts, the structure of which must resist deformation and destruction. The corresponding requirements are established in regulatory documents – see, for example, CFR Part 25, § 25.561, § 25.562 and § 25.785. The required strength of the structure is determined by the ultimate values of inertial loads, expressed, as in the previous two cases, through the maximum values of accelerations, $\{a_{max}\}$. Consequently, against the third hazard factor, one physical barrier is provided, i.e. $N_{pb}^{(3)} = 1$. The use of airbags was studied for military helicopters. They do not belong to physical barriers, but rather are one of the safety means that supplement the barrier function. The result of the study was the corresponding modernization of some types of helicopters (Jackson and Fasanella, 2003).

The fourth group of hazard factors – deformations of compartments occupied by people or cargo – is resisted by the fuselage structure of the aircraft. As already stated above, for the fuselage of a transport category airplane, impact with a hard landing surface is not a design case, however, among the regulatory provisions there is an indirect requirement for the structure to perform the function of the first physical barrier. In addition, seats and mass items (including their attachment units) must not deform under loads not exceeding the specified limits (expressed in terms of maximum acceleration values) in such a way that the deformations would impede rapid evacuation after emergency landing (CFR Part 25, § 25.561). Quantitative limitations of deformations, or relative displacements of parts of the structure, $\{u_{max}\}$, are not specified. The fuselage structure, structures of seats and mass items can be considered as three physical barriers preventing dangerous deformation, i.e. $N_{pb}^{(4)} = 3$. It should be noted, however, that only those deformations that can impede evacuation are considered as dangerous, and other threatening changes in form are not mentioned in the regulatory documents. As a result, there is uncertainty regarding the life volume of the cockpit or passenger cabin, as well as the possible damage to valuable or dangerous cargo when contacting the structural elements of the cargo compartment. This uncertainty is due to the a priori unknown behavior of the fuselage structure as the first physical barrier during an emergency landing on the ground.

The fifth hazard factor – the destruction of pipelines and tanks of an aircraft or transported cargo upon the landing impact (which can pose a threat of fire, explosion and environmental pollution) – is countered primarily by their structure, which must have an appropriate level of strength. Strength requirements are expressed, as in the previous cases, in terms of maximum acceleration values (CFR Part 25, § 25.561). If any pipelines or tanks can be damaged by mass items that were placed in compartments and torn off from their attachment units, then the structure of the corresponding compartment and the structure of the mass item, including the attachment units, act as potential additional barriers. Thus, the total number of physical barriers is in the range from one to three ($1 \le N_{pb}^{(5)} \le 3$). In the framework of the Aircraft Crashworthiness Research Program (ACRP), the crash resistant fuel systems of transport airplanes and helicopters were investigated and the fuel containment concepts were tested (Frings, 1998). The containment can serve as an additional physical barrier. As a result of this work, the Advisory Circular 29-2B used for certification of helicopters has been updated in provisions related to the fuel systems resistant to destruction in landing impacts[4].

The sixth hazard factor – destruction of fuselage elements, leading to the aircraft sinking – applies exclusively to the case of ditching (CFR Part 25, § 25.563 and § 25.801). The fuselage structure, including external doors and windows, acts as a physical barrier to prevent rapid sinking and to allow occupants to evacuate. Thus, $N_{pb}^{(6)} = 1$. Along with this hazard factor, in the case of ditching, other hazards discussed above can also come into play. The regulations provide for the investigation of behavior of the airplane "by model tests or by comparison with airplanes of similar configuration for which the ditching characteristics are known" (ibid, § 25.801).

The results of the assessment of the entire set of physical barriers that counteract all the hazard factors considered during an emergency landing can be formulated as follows:

- regulatory documents provide for such barriers in the aircraft design (although the term "physical barrier" is not used);
- the number of barriers, or levels of protection, varies from one to three depending on the hazard factor;
- these barriers are independent (in the sense that the failure of one does not lead to the failure of the other);
- the required properties of some barriers are not defined by regulatory provisions;
- the maximum acceleration values by which the requirements for other barriers are specified seem to be insufficiently substantiated when applied to new types of aircraft;
- as a result, the protection created by physical barriers is inferior in depth and reliability to the indicators typical for nuclear facilities;

[4]The current version of this document, updated in 2023, is 29-2C (AC 29-2C, 2023).

- it is inferior in a number of indicators to the capabilities created by the crashworthiness concept, which is used in the design of military helicopters;
- the protection currently provided for transport category airplanes using physical barriers cannot be considered sufficient, since it does not ensure safety to the extent that is achievable with the current state of science and technology.

7.2.3 Functional Levels of Protection

Let us consider the functional levels of protection against hazard factors in emergency landing, which are formed by technical and organizational measures. For the protection system that serves as our benchmark – defense in depth at nuclear facilities – they are presented in Sect. 1.1.5. It is obvious that the measures corresponding to the first level – prevention of deviations from normal operation conditions – are implemented at all stages of the aircraft life cycle. These measures include the use of reliable technologies for design, production, testing, maintenance and repair with high quality performance of all work tasks, certification of aircraft, licensing of operators (operating organizations and pilots) and compliance with operating rules.

The second functional level of protection – detection of deviations from normal operation (normal flight) conditions using special systems and devices and an attempt to take the deviations under control – is implemented by automatic flight control systems, such as autopilot or the thrust control system, as well as aerodynamic controls, landing gear control system, etc. In modern aircraft, these systems are included as components in the integrated flight control system (FCS). If FCS is equipped with AI elements, it is considered an intelligent flight control system, IFCS (Stengel, 1992; NASA, 2002). As soon as any of these systems identifies deviations from normal flight conditions, pilots receive warning signals and must take actions aimed at compensating for these deviations – see, for example, CFR Part 25, § 25.1585.

The third level should include measures taken when it is impossible to control deviations. These measures should prevent the development of an accident through the use of engineered safety systems and (or) the properties of inherent safety. But airworthiness standards do not define anything of the sort. The terms "safety systems" and "inherent safety" are not used in aviation. However, the ability of an aircraft to glide due to its aerodynamic quality in the event of engine shutdown can be considered a property of inherent safety, and this property is of course used. In any case, preventing the development of an accident is the responsibility of the pilot – his actions in emergency situations are described in the airplane flight manuals (AFM) and pilot's operating handbooks (POH) for particular airplane types. General recommendations can be found in Chapter 18 of the handbook FAA-H-8083-3C (2021). When the pilot decides to perform an emergency landing, his subsequent actions constitute the content of the pre-landing control phase. Thus, the third functional level of protection for transport category airplanes is not technically developed and is implemented by a human operator. Control before contact with the landing surface must ensure contact conditions that will exclude fatal consequences, minimize possible injuries to occupants and other damage (see Sect. 7.1.2). What these conditions represent in quantitative terms, and what the consequences of non-compliance with them might be, remain generally unclear. Note that in military aviation, when the pilot considers it impossible to make a "soft" emergency landing, he ejects. Ejection seats or crew escape capsules (Spirochkin, 2023) can be classified as safety systems (active safety systems).

The fourth level – managing an accident resulting from a failure of the previous level, using protective measures to mitigate its consequences and prevent a catastrophe – is associated primarily with the phase of main landing impact. In this phase, protection functions in the form of physical barriers, limiting the intensity of hazard factors to values that presumably exclude fatal consequences, reducing injuries and material damage (Sect. 7.2.2).

The fifth functional level of protection, as applied to emergency landing, can be interpreted as mitigation of the harmful consequences of the destruction of the aircraft, including injuries to occupants, sinking of the aircraft in the event of ditching, fire, explosion and environmental contamination. This level is implemented in the phase of post-impact processes through the use of intact physical barriers, structural elements and equipment that ensure the evacuation of passengers and crew members – see, for example, CFR Part 25, § 25.801, § 25.803, § 25.807, § 25.809 – § 25.813, § 25.815 and AC 25-17A (2016) – and through appropriate rescue measures.

A summary assessment of the functional levels of protection against hazard factors in emergency landing is as follows:

- such levels are not defined in regulatory documents, but their equivalents can be found among the provided technical and organizational measures;
- the number of levels that are independent can be considered equal to five, which corresponds to the depth of protection of nuclear facilities;
- the third level – measures taken when it is impossible to control deviations from normal flight conditions and preventing the development of an accident – is implemented, in contrast to the protection of nuclear facilities, not through the engineered safety systems or the properties of inherent safety, but through the actions of the human operator (pilot) in the pre-landing control phase;
- the purpose of this control is to ensure emergency landing conditions that would exclude fatal consequences, minimize possible injuries to occupants and other damage, but the quantitative relationship between landing conditions and the values of hazard factors during the main landing impact is not determined;
- this uncertainty seriously complicates predictive safety assessment and accident management;
- in military aviation, in the event of failure of the third level of protection, the rescue of crew members in ejection seats or crew escape capsules is carried out, but such measures are not provided for transport category airplanes;

- due to the weakness of the third functional level, caused by the above-mentioned uncertainty, the unreliability of human factors (embodied in the pilot) and the absence of automatic safety systems, the protection provided cannot be considered sufficient, especially taking into account the modern scientific and technical potential.

7.2.4 Assessment Totals

The assessments given in Sects. 7.2.2 and 7.2.3 demonstrate certain logic in construction and relative completeness of the protection system formed by the requirements of regulatory documents regarding emergency landings of transport category airplanes. Fulfilment of these requirements at the design stage ensures safety within the limits considered acceptable by regulators. However, this protection is weaker in depth and reliability as compared with the defense in depth system used at nuclear facilities, which we recognize as the most advanced. Moreover, in a number of respects it is inferior to the crashworthiness concept applied in military aviation. For transport category airplanes, the functional level of protection preventing the development of an accident using the engineered safety systems or the properties of inherent safety is poorly developed. It is implemented in the pre-landing control phase due to the actions of the pilots. But the relationship between the emergency landing conditions (including the properties of the chosen landing surface and the aircraft kinematic parameters) that can be achieved by such control and the values of the hazard factors that are realized during the main landing impact is not quantitatively determined. It may exist in some form in the heads of designers or pilots, based on their professional skills. But this form is rather vague and not quantified.

The uncertainty of quantitative relationships between operating conditions causing extreme loads and the parameters of the system's response to them is apparently generally inherent in emergency situations that develop into accidents, accompanied by nonlinear mechanical processes, including partial destruction of the system. Due to the complexity of analyzing such processes, the complexity of the system itself, and the relatively infrequent occurrence of accidents, many of their types remain insufficiently studied. The consequence of this is uncertainty about safety in the event of a possible accident. The specification of maximum acceleration values for the case of emergency landing by regulatory documents can be considered as an attempt to solve this problem. However, it is not enough to limit only the response parameters that are hazard factors without relating them to the emergency landing conditions and other significant variables. It is necessary to know the relevant relationship in a detailed quantitative form in order to guarantee a certain level of safety by controlling the key variables. This possibility is provided by the current level of development of theoretical and experimental research methods, but it is not implemented in the design of transport category airplanes.

The required properties of some physical barriers in the path of hazard factors are not determined in regulatory documents or are not specified reliably enough. This makes it difficult to forecast the level of safety and ensure it in practice in the event of emergency landing.

The level of safety is negatively affected by the absence of active safety systems (including automatic ones), limited use of passive safety devices and the potential unreliability of human factors, embodied in the pilot, who is subject to significant psychological pressure during an accident. Current approaches to the design of transport category airplanes are inconsistent with an important principle embodied in the defense in depth system at nuclear facilities, which is established in the international safety requirements for the design of nuclear power plants (IAEA, 2016) and reproduced in Sect. 1.1.5. According to this principle, the need for rapid intervention by a human operator in emergency management must be reduced to minimum. Alternatively, it is acceptable to demonstrate that the operator has sufficient time to make a decision and take corrective action (ibid). However, given the high speed of processes accompanying an emergency landing, it is questionable whether such an alternative can be relied upon in any case.

The instrumentation currently used on transport category airplanes does not provide the ability to obtain all the information needed to determine the actual state of the aircraft and to enable the pilot to make informed decisions on accident management. In particular, the pilot or the automatic control system does not have the data characterizing the actual properties of the structural elements that perform the function of physical barriers, as well as the properties of the potential landing surface, which are necessary for forecasting the behavior of the structure and the level of safety in emergency landing conditions and choosing the most acceptable combination of these conditions.

In addition, the instrumentation available on board does not allow for a full analysis of the aircraft state after the accident. This is not of decisive importance if the accident occurred in a populated and industrialized region, where such a service could be provided by a nearby specialized company. However, the instrumentation for the analysis in question is necessary in the case of a mission conducted in uninhabited areas or at a distance from Earth (in the event of a spacecraft or aerospace vehicle accident).

The results of the assessment of the existing system of protection against hazard factors in emergency landing can be formulated as follows:

- this protection is characterized by a small number of physical barriers, uncertainty of the properties of some of them and insufficiently substantiated requirements for others, uncertainty of the relationship between accident conditions and hazard factors, and underdevelopment of safety systems;
- all of the above indicates that safety is ensured formally, not in essence, resulting in a safety deficit;
- however, the safety deficit is not insurmountable – it can be minimized through R&D that must be in line with the current level of science and technology.

7.3 Potential of the SPARS Concept to Minimize Uncertainty and Safety Deficit in Emergency Landing

7.3.1 General Considerations for Applying the SPARS Concept to the Design Case "Emergency Landing"

The shortcomings of the existing system of protection against hazard factors in emergency landings of transport category airplanes, identified in the previous section, can be eliminated within the framework of the SPARS concept applicable in the design of new aircraft, including aerospace vehicles. Below, its capabilities to minimize uncertainty and improve safety in this emergency situation are presented. They enhance the capabilities of traditional design approaches. The SPARS concept incorporates the methods to ensure safety corresponding to these approaches and supplements them with a number of innovations (Sect. 6.4.3).

Considering the diversity of possible causes of emergency landings, as well as states of the aircraft structure and systems, the variability of the pilot's actions, the wide range of landing conditions, as well as the insufficient study of nonlinear processes of deformation and destruction of the structure under impact loads, it is obvious that at the design stage it is difficult to foresee all scenarios for the development of such accident and assess their outcomes. In this case, designers are dealing with a combination of aleatory and epistemic uncertainty. The application of the SPARS concept is aimed at minimizing uncertainty and compensating for the safety deficit by giving the aircraft properties of "smartness", "proactivity" and "resilience" to assess emerging dangers, including previously unknown ones, in real time and respond to them with minimal negative consequences.

An emergency landing can be considered as the result of the transformation of a critical situation that arose in flight, which could not be prevented, into an accident, which in turn threatens to develop into a catastrophe. The capabilities of the SPARS concept to manage this transformation and prevent catastrophic development are realized through:

- determining the quantitative relationship between the conditions characterizing an emergency landing and the hazard factors in this case;
- improving the physical barriers that counteract these factors;
- using safety systems and other safety means;
- using the Critical and Emergency Control System (CECS).

The set of hazard factors that arise in an emergency landing, $\mathbb{H} = (\{a_{max}\}, \{u_{max}\}, ...)$, can be expressed as a generalized function of a number of arguments:

$$\mathbb{H} = \mathcal{F}(\mathbb{S}, \mathbb{P}, \{V\}, \{\Psi\}, \mathfrak{S}, \mathfrak{H}, \mathfrak{A}), \tag{7.3}$$

where $\mathbb{S}$ is the set of characteristics of the aircraft structure and its cargo, including configuration, dimensions, mechanical properties of structural materials, etc.;

the $\mathbb{P}$ set covers the landing surface parameters that affect the behavior of the aircraft structure when it impacts that surface: relief data in geographic coordinates

X and Y, $\mathcal{R}(X, Y)$; mechanical properties of soil, concrete, etc. (density ρ, Young's modulus E, shear modulus G, kinematic viscosity ν, ultimate strength σ, friction coefficient f_t, etc.), i.e. $\mathbb{P} = (\mathcal{R}(X, Y), \rho, E, G, \nu, \sigma, f_t, \ldots)$;

$\{V\} = \{V_x, V_y, V_z,\}$ is the velocity vector at the moment of touching the landing surface;

$\{\Psi\} = \{\psi, \vartheta, \gamma\}$ is a column vector containing the rotation angles of the aircraft axes: yaw angle ψ, pitch angle ϑ and roll angle γ;

the arguments $\mathfrak{S}$, $\mathfrak{H}$, and $\mathfrak{A}$ describe the properties and functions of safety systems and other safety means if they are used, the influence of human factors (including the pilot performing control) and actions of automatic control, respectively.

Specifications for design must contain allowable values for all hazard factors, $\mathbb{H}_a$. Therefore, having $\mathcal{F}$, for known argument values, it is possible to evaluate hazard factors $\mathbb{H}$ and check the fulfillment of safety criteria:

$$\mathbb{H} \leq \mathbb{H}_a. \tag{7.4}$$

On the other hand, based on $\mathbb{H}_a$ it is possible for known parameters of the aircraft structure, $\mathbb{S}$, and known or expected values of some arguments – $\mathbb{P}$, $\{V\}$, etc., to determine the required values of other arguments – properties and functions of safety systems and other safety means, $\mathfrak{S}$, or actions of automatic control, $\mathfrak{A}$, that could ensure safety in the event of an emergency landing.

The generalized function $\mathcal{F}$ includes a set of dependencies expressed through analytical formulas, tables, algorithms associated with a numerical model, or other mathematical objects. The possibilities of obtaining the dependencies of hazard factors on the arguments $\mathbb{S}$, $\mathbb{P}$, $\{V\}$ and $\{\Psi\}$ are considered in Sect. 7.4.

The set $\mathbb{S}$ includes as a subset the characteristics of physical barriers, $\mathbb{S}_{pb}$. The type of technical system limits the possible number of such barriers. For aircraft, it is problematic to increase it to values achievable in NPPs or other ground facilities. Additional barriers make the aircraft heavier, and this may be unacceptable according to the criteria of weight optimization. The possibilities provided by the SPARS concept in this regard are discussed below in Sect. 7.3.2.

The insufficient depth of protection due to the limited number of physical barriers can be compensated for by using safety systems and other safety means. Their parameters, expressed in Eq. (7.3) through the argument $\mathfrak{S}$, are capable of exerting a significant positive influence on safety in emergency landing. Section 7.3.3 introduces some ideas being developed in this direction within the SPARS concept.

The main design measure for implementing the properties of "smartness", "proactivity" and "resilience" in an aircraft is equipping it with a special automatic control system – CECS (see above), capable of recognizing a critical situation in flight, preventing it from turning into emergency landing by appropriate control, and in the event of failure of that control, ensuring survival and minimizing damage. CECS should solve these tasks in real time. A significant portion of the necessary equipment (sensors, communications, processing devices and actuators) is already used in modern advanced airplanes. The SPARS concept envisages the inclusion of CECS into FCS or IFCS (variants of flight control system mentioned in Sect. 7.2.3) and its interaction with other components of these systems, including avionics.

If necessary, CECS should minimize the pilot's involvement in emergency management. His involvement relates mainly to the pre-landing control phase and it is a potential source of uncertainty that can negatively affect the safety in emergency landing – as the argument $\mathfrak{H}$ in Eq. (7.3). The functions of CECS and the corresponding automatic actions, described in this equation by the argument $\mathfrak{A}$, are presented in Sect. 7.3.4.

7.3.2 Improving the Physical Barriers

As shown in Sect. 7.2.2, the hazard factors in an emergency landing of a transport category airplane are counteracted by a relatively small number of physical barriers. The regulatory provisions leave the properties of some of them undetermined, while the requirements for others, specified by maximum acceleration values, seem to be insufficiently substantiated.

In particular, the required reliability of the fuselage structure as the first physical barrier upon impact with a hard landing surface is uncertain. To minimize uncertainty and ensure safety in this case, the SPARS concept includes improvements to the fuselage structure. Its lower part, which comes into contact with the landing surface, must have the properties of controlled (programmable) destruction and absorption of a certain amount of kinetic energy. Deformations of other parts of the fuselage must be within such limits that during an emergency landing the life volume of the cockpit or passenger cabin is preserved and valuable or dangerous cargo (payload) is not damaged due to not allowable contacts with structural elements of the cargo compartment (payload bay). These requirements correspond to the principles of crashworthiness incorporated into the SPARS concept. Their fulfillment must be confirmed by calculations and experiments at the design stage of the aircraft.

The SPARS concept involves substantiating the maximum values of accelerations, $\{a_{max}\}$, which determine the required properties of other physical barriers including (see Sect. 7.2.2):

- passenger or crew member seat with restraining belts;
- cargo (payload) and equipment attachment units;
- cargo compartment (payload bay) structure;
- restraining structural elements that prevent movements of the torn off mass items;
- structures of pipelines and tanks of the aircraft or cargo (taking into account the threat of fire, explosion and environmental pollution in the event of their destruction during the landing impact).

As part of improving the specification of maximum acceleration values, the SPARS concept provides for considering the influence of masses of large-sized and heavy payloads, as well as their structural stiffness parameters, on the dynamic response of the aircraft structure during emergency landing. When such influence is significant, the requirements for the load-bearing capacity of attachment units and structural elements of the payload cannot be established using a priory known values of maximum acceleration values at its center of gravity. The strength of these units and structural elements must be substantiated by a dynamic analysis of the mechanical system including the aircraft and the payload, by an appropriate experiment, or another equivalent procedure covering the entire system.

The complete set of requirements for the resistance of physical barriers must be expressed through the allowable values of the hazard factors, $\mathbb{H}_a$. Measures for improving the physical barriers are based on obtaining a generalized function $\mathcal{F}$, expressing the dependencies between the conditions of emergency landing and the hazard factors (see Sect. 7.3.1). Being known, these dependencies allow designers to determine the characteristics of the barriers, $\mathbb{S}_{pb}$, at which the hazard factors do not exceed the allowable values with the expected values of other arguments of the function $\mathcal{F}$.

In Sect. 7.3.1, the problem of increasing the number of physical barriers due to the limitations imposed by weight optimization was mentioned. For example, adding a second, internal shell to the fuselage structure, thus creating a passenger or cargo compartment in the form of a capsule inserted into the outer shell would help improve safety in the event of emergency landing. But the price for this measure would be an increase in the mass of the aircraft, with a simultaneous increase in the required engine power and fuel consumption per flight. Detailed studies are needed to substantiate the effectiveness of such measures. Given such limitations, a protective shell for fuel system components seems to be the only unconditional candidate for the role of an additional physical barrier. It can be comparatively compact and light. Studies in this direction conducted earlier (Frings, 1998) deserve to be continued within the framework of the SPARS concept.

A reasonable alternative to installing additional barriers is to improve the properties of existing ones. In particular, the resistance of structural elements to impact loads can be increased by manufacturing them from metamaterials that can change their physical and mechanical properties under high-speed loading in such a way as to absorb significant kinetic energy at small relative displacements – see, for example, (Synbiosys, 2021; Liang and Crosby, 2022).

7.3.3 Safety Systems and Other Safety Means

In conditions where the number of physical barriers against the hazard factors in emergency landing is small, and the load-bearing capacity of the structure is only substantiated for normal operating modes, protective functions aimed at minimizing harm can be performed by safety systems and other safety means. The SPARS concept envisages equipping an aircraft with safety systems whose functions are similar to those used at NPPs (NP-001-15): a *controlling safety system*, a *protecting safety system* and a *localizing safety system*.

The controlling safety system is activated at the end of the pre-landing control phase when CECS determines that some of the emergency landing conditions (properties of the landing surface, the aircraft velocity vector or the rotation angles of the aircraft axes) do not ensure the required level of safety (corresponding to the allowable values of the hazard factors, $\mathbb{H}_a$). Due to the short time for decision-making and in order to minimize the uncertainty associated with the actions (or inactions) of the pilot, the activation and functions of the controlling safety system must be performed automatically. To increase safety in emergency landing, it may be necessary to avoid the aircraft colliding with random obstacles on the landing surface, further reduce the vertical velocity, or change one of the rotation angles. When control by aerodynamic surfaces cannot be relied upon, small solid-fuel rocket engines installed in certain places of the aircraft can be used as actuators of the controlling safety system. Their prototypes can be those used for ejection of pilots of military aircraft and helicopters or for ensuring soft landings of the Soviet reentry ballistic-type capsules *Vostok*, *Voskhod* and *Soyuz*.

The protecting safety system prevents uncontrolled movements of occupants or torn off mass items during the main landing impact and post-impact processes. It must come into action if the control in the first phase of an emergency landing, including work of the controlling safety system, failed, as well as in the event of failure of physical barriers such as seats with restraining belts, attachment units of the mass items, or structural elements of the cargo compartment that prevent the movement of the torn off items. The action is initiated by CECS, which identifies pre-landing control failure and (or) barrier failures. A possible implementation for the protecting safety system is to fill a part of the volume of the cockpit, as well as the passenger or cargo compartments, with foam, which could quickly acquire elastic properties (IdeaConnection, 2015). This system can help maintain the life volume of the compartments occupied by people, and also perform a localization function, stopping the movement of the torn off cargo within the cargo compartment.

In general, the localization function is associated with limiting the spread of flammable, toxic or radioactive substances released from destroyed pipelines and tanks. This function is performed by the localizing safety system, which starts working on a command of CECS and compensates for the failure of the physical barrier formed by the structure of the pipeline or tank containing the hazardous substance, or, if the protective shell is used (see Sect. 7.3.2), for the destruction of such a shell.

All three safety systems considered are active safety systems in the sense that they function when energy is supplied to them from an artificial source. Initiation of each of them is part of the automatic control actions, $\mathfrak{A}$, and is provided by CECS.

Along with active safety systems, passive safety means (or devices) can also be used, which operate by natural actuation, without artificial energy sources, e.g., under the inertial forces. The range of passive safety devices include programmable, destructible elements built into the load-bearing structure that limit the loads transmitted to other critical parts, such as passenger or pilot seats. Elements of this kind comply with the crashworthiness principles integrated into the SPARS concept.

The development of safety systems and other safety means is an important area of R&D in the design of advanced airplanes and aerospace vehicles and one of the distinguishing features of the SPARS concept. The required properties and functions of these systems and means, $\mathfrak{S}$, must be determined based on the allowable values of the hazard factors, $\mathbb{H}_a$, and the known generalized function $\mathcal{F}$ with given – achievable or expected – values of other arguments of this function (Sect. 7.3.1).

7.3.4 Critical and Emergency Control System (CECS)

The system described below and abbreviated as CECS, is intended to support the process of aircraft control, taking into account the possibility of a critical or emergency situation occurring in flight, which can develop into an accident. Some of the CECS functions are implemented under normal flight conditions in order to recognize early signs of a critical situation and impending adverse developments, i.e. to prevent an accident. If preventative measures fail, then other functions come into play to ensure accident management – as far as this is possible. CECS uses sensors, communications and instruments that provide in-service monitoring of the aircraft flight conditions, the health of its structure and performance of its systems, and diagnosing anomalies in real time. As noted in Sect. 7.3.1, some equipment necessary for this purpose is already available in modern advanced airplanes. The use of this equipment within the SPARS concept and the interaction of CECS with other components of IFCS are complemented by a number of innovations.

Part of these innovations consists of installing additional sensors to expand the range of parameters recorded in order to meet the requirements specific to the SPARS concept, which are listed under numbers 1 and 2 in Sect. 6.4.1. Innovations are also being introduced in the processing of information received from sensors to meet other requirements listed there, in particular 3 and 5. For example, processing compensates for fragmentation caused by gaps in data due to the absence of sensors at certain points on the aircraft or large time intervals between successive sensor records. Compensation is achieved by reconstructing the fields of variables describing the environment surrounding the aircraft and the state of its structure and systems. Such reconstruction meets the ideas of information support for the life cycle of technical systems. This makes it possible to minimize the uncertainty associated with missing data and to recognize subtle signs of change that could lead to potentially dangerous deviations from normal operating conditions, as well as to detect hidden deviations. Computational and logical operations for reconstruction and recognition, identification of a critical or emergency situation, determination of control actions and analysis of their effectiveness are carried out using electronic devices belonging to CECS. To implement a number of intellectual functions the neural network technology is envisaged. For small changes that have not yet reached the values that would characterize them as deviations, proactive (hazard-preventing) control is carried out – it corresponds to the subset of actions $\mathfrak{A}_p$. Reactive (responding to the emerging danger) control is applied to deviations – in the form of actions $\mathfrak{A}_r$. Both subsets belong to the argument $\mathfrak{A}$ of the generalized function $\mathcal{F}$ (Sect. 7.3.1):

$$\mathfrak{A}_p \in \mathfrak{A}, \quad \mathfrak{A}_r \in \mathfrak{A}. \tag{7.5}$$

According to the SPARS concept, both proactive and reactive (compensatory) actions should be carried out predominantly in automatic mode.

In the event of an emergency landing that could not be prevented, CECS determines in real time control actions capable of minimizing the expected values of hazard factors (and thus the possible harmful consequences). They are also covered by the argument $\mathfrak{A}$ of the generalized function $\mathcal{F}$. In particular, in the pre-landing control phase, based on the forecast of the behavior of the aircraft structure using this function and the allowable values of the hazard factors, $\mathbb{H}_a$, the required properties of the landing surface, $\mathbb{P}$, and the kinematic parameters of the aircraft, $\{V\}$ and $\{\Psi\}$, are being found, which ensure that the allowable values are not exceeded. The required $\mathbb{P}$ values determined in this way enables the choice of an emergency landing site. This choice is being made on the basis of terrain information loaded into the Terrain Awareness and Warning System (TAWS) (FAA-H-8083-6, 2009) and supplemented by data on the mechanical properties of the soil in the vicinity in its current state[5]. Based on the chosen emergency landing site (surface) and the required values of $\{V\}$ and $\{\Psi\}$, automatic control actions, $\mathfrak{A}$, can be determined to ensure flight to this site and acceptable landing conditions. When choosing the emergency landing site, it is also necessary to take into account the state of the aircraft structure and systems, the availability of nearby airfields (where the rescue of people is facilitated by the presence of personnel, technical means and infrastructure) and weather conditions (wind, precipitation, visibility, the possibility of icing). The methodology for planning an emergency landing can be found, for example, in the publication by Meuleau et al. (2009a). It is implemented by the emergency landing planner (Meuleau et al., 2009b), a dedicated component of IFCS.

In the event of failure of the pre-landing control, i.e. the impossibility of choosing the values of $\mathbb{P}$, $\{V\}$ or $\{\Psi\}$ that ensure acceptable levels of hazard factors, the safety systems are activated (Sect. 7.3.3). The controlling safety system is activated at the end of the first phase of emergency landing, and the protecting and localizing safety systems come into action in the second and third phases.

In the main landing impact phase, CECS monitors the work of physical barriers and safety systems, records data on the behavior of the aircraft structure and systems, as well as hazard factors. If necessary, the localizing safety system

[5] The current state implies taking into account seasonal and average periodic meteorological variations.

is activated. Monitoring and recording continue during the post-impact processes phase.

Some functions of CECS are also implemented after emergency landing. These primarily include monitoring the work of the localizing safety system, as well as analyzing the aircraft state based on data received from sensors. Another function is storing the accident data to expand the training set that the neural network will use in similar situations in the future.

Table 7.2 contains a list of operations performed by CECS under normal flight conditions and during an emergency landing. These operations are aligned with the functional levels of protection against hazard factors considered in Sect. 7.2.3.

Among the many operations performed by CECS, an important one is the forecast of the behavior of the aircraft structure during emergency landing (operation 8). The forecast is based on known structural characteristics, $\mathbb{S}$, as well as input data on the properties of the landing surface, $\mathbb{P}$, and the kinematic parameters at the moment of its contact, $\{V\}$ and $\{\Psi\}$. The final result of the forecast is the values of the hazard factors, $\mathbb{H}$. By comparing them with the specified values of $\mathbb{H}_a$, if the criteria expressed by Eq. (7.4) are met, a conclusion can be made about a sufficient level of safety. Knowledge of the quantitative relationship between the hazard factors, on the one hand, and the variables covered by the arguments $\mathbb{S}$, $\mathbb{P}$, $\{V\}$, $\{\Psi\}$, etc., on the other hand, i.e. knowledge of the generalized function $\mathcal{F}$, makes it possible to choose relatively safe emergency landing conditions for a certain aircraft. Obtaining the dependencies expressed by this function is one of the key aspects of the methodology to ensure safety within the SPARS concept. Section 7.4 describes the determination of the dependencies in question using the example of a possible emergency landing of the *Buran* aerospace vehicle.

7.4 Determination of Structural Behavior and Safety Conditions in the Postulated Emergency Landing of the *Buran* Aerospace Vehicle

7.4.1 Dynamic analysis

The behavior of the aircraft structure in an emergency landing can be determined by dynamic analysis using an appropriate mathematical model. This approach was implemented during the design of the Soviet space transport system *Energia-Buran* (see Figure 6.8) in the 1980s. The methodology of dynamic analysis applicable to the emergency landing of the *Buran* aerospace vehicle – a hypothetical case, but postulated in the design specifications – was absent at that time, and it was developed by the author (Spirochkin, 1987).

The analysis was based on the numerical solution of differential equations describing:

- the movement of the aircraft as a rigid body during impact with the landing surface;
- vibrations of its structure, payload and equipment;

Table 7.2 Operations performed by CECS in relation to emergency landing.

Flight conditions	*Level of protection against hazard factors*	*Operations performed by CECS*
Normal flight	1. Preventing deviations from normal operation conditions (normal flight)	1. Identification of changes occurring outside and inside the aircraft that can cause deviations, under limited recorded data. 2. Assessing the degree of danger of these changes, taking into account the environment and the current state of the aircraft. 3. If there is a potential danger, determination of proactive control actions, $\mathfrak{A}_p$.
Critical situation (potentially resulting in emergency landing)	2. Detection of deviations from normal flight using special systems and devices and an attempt to take the deviations under control	4. Identification of deviations, including hidden ones, in case of failure of proactive control, under limited recorded data. 5. Determination of reactive control actions, $\mathfrak{A}_r$, to compensate for deviations taking into account the environment and the current state of the aircraft. 6. Assessing the effectiveness of these actions and, in case of ineffectiveness, identification of an emergency situation.

Contd...

Table 7.2 *Contd...*

Flight conditions	*Level of protection against hazard factors*	*Operations performed by CECS*
	Emergency landing	
• The pre-landing control phase	3. Measures taken when it is impossible to control deviations that prevent the development of an accident through the use of engineered safety systems and (or) the properties of inherent safety	7. Analysis of possible emergency landing options and choice of a suitable landing site (surface), i.e. determination of the set $\mathbb{P}$. 8. Forecast of the behavior of the aircraft structure under possible kinematic parameters at the moment of contact with the landing surface and choice of the values of $\{V\}$ and $\{\Psi\}$ that ensure an acceptable level of safety. 9. Based on the choice of $\mathbb{P}$, $\{V\}$ and $\{\Psi\}$, constructing the trajectory of the aircraft to the emergency landing site and determining the required control actions $\mathfrak{A}$. 10. In case of failure of pre-landing control, activation of the controlling safety system.
• The main landing impact phase	4. Managing the accident using protective measures to mitigate its consequences and prevent a catastrophe	11. Activation of the protecting safety system and, if necessary, localizing safety systems . 12. Monitoring the work of physical barriers and safety systems. 13. Recording data on the behavior of the aircraft structure and systems, as well as hazard factors.
• The post-impact processes phase	5. Mitigation of the harmful consequences of aircraft destruction	Continuation of the level 4 operations, plus: 14. Monitoring the ongoing work of the localizing safety system.
After emergency landing	Continuation of the level 5 function	If necessary, operation 14, and in addition: 15. Post-accident analysis of the aircraft state. 16. Storing the accident data to expand the training set for the neural network (for use in similar situations in the future).

- large plastic strains, changes in the shape of structural elements (due to large relative displacements) and their destruction under impact loads;
- mechanical interaction of the fuselage with the landing surface.

These equations were obtained by mathematical modeling of a dynamic system that included the aircraft, the mass items attached to it (payload and equipment), and the solid massif of concrete or soil forming the landing surface. The modeling was performed using FEM in a physically and geometrically nonlinear formulation.

To describe large nonlinearities, structural dynamics of this system was described in incremental form, meaning the unknowns were the increments of the kinematic parameters, $\Delta\{U\}_{t,\,t+\Delta t}$, $\Delta\{\dot{U}\}_{t,\,t+\Delta t}$, and $\Delta\{\ddot{U}\}_{t,\,t+\Delta t}$, over a small time interval $(t, t+\Delta t)$:

$$[M]_{t,\,t+\Delta t}\,\Delta\{\ddot{U}\}_{t,\,t+\Delta t} + [C]_{t,\,t+\Delta t}\,\Delta\{\dot{U}\}_{t,\,t+\Delta t} + [K]_{t,\,t+\Delta t}\,\Delta\{U\}_{t,\,t+\Delta t} =$$
$$= \{F(t+\Delta t, \{U\}_{t+\Delta t}, \{\dot{U}\}_{t+\Delta t} \ldots)\} - \{F(t, \{U\}_t, \{\dot{U}\}_t \ldots)\}, \qquad (7.6)$$

where $[M]_{t,\,t+\Delta t}$, $[C]_{t,\,t+\Delta t}$ and $[K]_{t,\,t+\Delta t}$ are matrices of secant coefficients of inertia, damping and stiffness of the finite element ensemble, respectively, determined for this interval; the column vector F includes the forces acting on the finite element ensemble, the nature of which differs from the forces of elasticity, inertia and damping.

A comparison of this matrix equation with Eq. (4.22), which describes linear dynamic problems, demonstrates the significant complexity of Eq. (7.6). This complexity is due to the consideration of large displacements and strains of structural elements, changes in their shape and destruction under conditions of plastic flow and loss of stability, as well as variable conditions of contact of the fuselage with the landing surface in the presence of friction. To solve Eq. (7.6), numerical integration over time was used with the application of the implicit methods described in Sect. 4.2.3, which led to a series of nonlinear algebraic equations of the form:

$$[\hat{K}]_{t,t+\Delta t}\Delta\{U\}_{t,t+\Delta t} = \Delta\{\hat{F}\}_{t,t+\Delta t}, \qquad (7.7)$$

where $[\hat{K}]_{t,t+\Delta t}$ is the effective secant stiffness matrix;

$\Delta\{\hat{F}\}_{t,t+\Delta t}$ is the column vector of effective load increments.

These equations were solved by an iterative method, as a result of which the increments of nodal displacements, $\Delta\{U\}_{t,\,t+\Delta t}$, were determined. Based on these increments, nodal displacements, as well as nodal velocities and accelerations, were being obtained at the end of each time interval under consideration:

$$\{U\}_{t+\Delta t} = \{U\}_t + \Delta\{U\}_{t,t+\Delta t};$$
$$\{\dot{U}\}_{t+\Delta t} = \{\dot{U}\}_t + \Delta\{\dot{U}\}_{t,t+\Delta t}; \qquad (7.8)$$
$$\{\ddot{U}\}_{t+\Delta t} = \{\ddot{U}\}_t + \Delta\{\ddot{U}\}_{t,t+\Delta t}.$$

Along with the kinematic parameters, strains and stresses at points of the structure were calculated. Such calculations were being carried out at each loop of the iteration cycle aimed at solving Eq. (7.7) for elements of the dynamic system whose mechanical behavior was characterized as nonlinear. The methodology for solving the dynamic problem under consideration is described in detail in the author's dissertation (Spirochkin, 1987).

This methodology took into account the possibility of performing emergency landing on various landing surfaces with different mechanical properties: a concrete runway or soil with different ultimate strength values. The contact of the fuselage with the landing surface was simulated using special finite elements. For the case of landing on soil, their properties were assigned on the basis of experimental data accumulated during the experimental tests of landings of the Soviet reentry ballistic-type capsules. To study the features of the mechanical interaction between the bottom of the *Buran* fuselage (which has a low curvature – see Figure 7.1) and the landing surface under conditions of a sliding impact – in order to describe these features in the mathematical model – a special experimental facility was proposed (Spirochkin, 1988; Patrushev et al., 1989).

The mathematical model used to analyze the *Buran* emergency landing had about 1800 degrees of freedom, which seems like a very modest number today, but was a formidable task for the computers of the time. Possible modeling errors caused by a coarse finite element mesh were compensated by refining the model parameters based on experimental data. The model was validated by comparing the dynamic characteristics of the aerospace vehicle (natural frequencies and vibration mode shapes), obtained by solving the eigenvalue problem for this model, with the results of horizontal frequency tests of the *Buran* flight specimen, in the cargo compartment of which the unit of auxiliary devices was located (Figure 7.2). The construction of the model, calculation of dynamic characteristics and determination of the dynamic response of the structure in the main landing impact phase were carried out using a software system developed by the author and his colleagues, which was subsequently named *NewTone* (Spirochkin, 1993).

To determine the behavior of the *Buran* structure and safety conditions for various emergency landing scenarios, many dynamic analyses (computer simulations) were performed corresponding to different values of the properties of the landing surface and the kinematic parameters of the aerospace vehicle. The kinematic parameters corresponded to cases of symmetrical landing, i.e. without sideward movement of the vehicle as a rigid body (along the z axis). In each dynamic analysis, the relative linear displacements of the elements of the spatial structure of the vehicle in the directions of three axes: x, y, z and rotations around them were considered. A number of computer simulations were carried out for different payloads of varying sizes and masses. Each analysis yielded time histories for the following dynamic response parameters:

- components of the acceleration vectors at the characteristic points of the structure, as well as at the equipment and payload points, $\{a(t)\} = \{a_{1x}(t),$

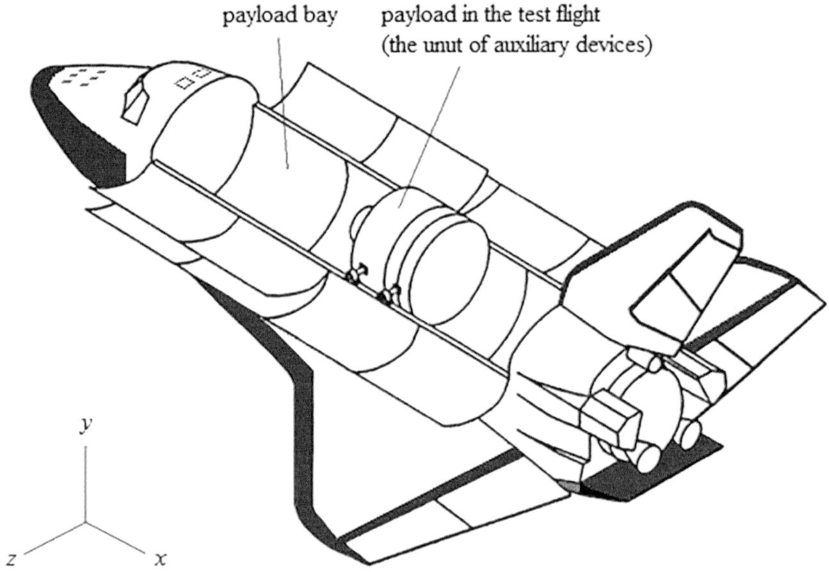

Fig. 7.2 *Buran* with its payload in the test flight - the unit of auxiliary devices.

$a_{1y}(t), a_{1z}(t), a_{2x}(t), \ldots a_{nz}(t)\}$[6], which made it possible to evaluate such a group of hazard factors as $\{a_{max}\}$;

- internal forces in the attachment units of payloads with large dimensions and mass, on the basis of which the possibility of their destruction was assessed;
- relative lateral displacements at points in the payload bay, $\{u_{1z}(t), u_{2z}(t), \ldots\}$, to estimate deformations of its structure, $\{u_{max}\}$, another group of hazard factors.

As an example, Figure 7.3 shows the calculated time histories of accelerations for the case of emergency landing of *Buran* on the soil runway with the maximum allowable vertical velocity.

The design requirements for the strength of the attachment units for the *Buran* payload and equipment were essentially the same as those in use at that time for transport category airplanes (ENLG, 1985): these units must withstand loads (inertial forces) applied at the center of gravity of each mass item that correspond to the following acceleration ranges:

- for longitudinal load – from zero to 9g when load acts forward and from zero to 1.5g when load acts backward;
- for vertical load – from zero to 4g when load acts downwards and from zero to 2g when load acts upwards;
- for sideward load – from -2.25g to +2.25g.

For payloads weighing 20 t or more, the maximum specified longitudinal load acting forward was reduced to a value corresponding to 6g (similar specifications were used in the design of the American *Space Shuttle* system). The comparison

[6]The variable n used as a subscript denotes the number of observed points.

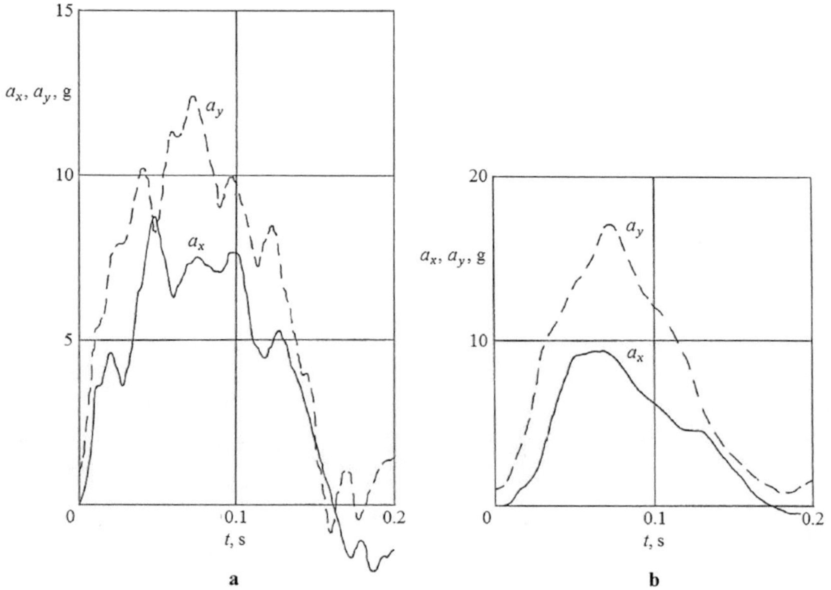

Fig. 7.3 Time histories of longitudinal (a_x) and vertical (a_y) accelerations for the case of emergency landing of *Buran* on the soil runway (ultimate strength $\sigma = 1.1$ MPa) with a vertical velocity $V_y = -3$ m/s: **a** on the fuselage longeron; **b** at the center of gravity of the payload (the unit of auxiliary devices).

with the data given in Eq. (7.2) shows some differences between the regulations of that time and the requirements of modern airworthiness standards. However, these differences are not fundamental.

Calculations carried out during the design of the *Energia-Buran* system led to the conclusion that loads on the attachment units of mass items do not exceed the specified limits if an emergency landing of *Buran* is performed on:

- the low-strength soil runway ($\sigma \leq 0.4$ MPa) with vertical velocity (at the moment of touch) approximately $V_y = -1$ m/s or
- the high-strength (e.g., frozen) soil or concrete runway (3.5 MPa $\leq \sigma \leq$ 8 MPa) with vertical velocity about $V_y = -0.7$ m/s.

The pitch angle (close to the angle of attack) must be maintained near the minimum allowable value ($\vartheta = 9°$).

Computer simulations enabled analysis engineers involved in the design to determine the influence of the dynamic properties of large-sized and heavy payloads on the forecast of behavior of the *Buran* structure and safety conditions during its emergency landing. Errors in estimating the dynamic response of such mass items through acceleration shock spectra (a common tool used by designers) were studied, as well as the impact of the stochastic variation of the load-bearing capacity of collapsing structural elements on the dynamic response and safety assessment.

The developed methodology of dynamic analysis made it possible to:

- consider at the design stage of an aerospace vehicle a set of hazard factors that act during an emergency landing (and not just the accelerations specified in regulatory documents);
- determine loads on equipment, payload and pilot seats, taking into account their placement, attachment features and partial dynamic characteristics, as well as structural parameters and dynamics of the vehicle itself (which represents an evolution of both regulatory provisions and traditional design practice);
- assess how landing conditions that can be controlled in the time interval preceding the main landing impact affect loads and other hazard factors (which creates potential for emergency management).

Thus, this methodology allows designers to determine a generalized function $\mathcal{F}$ that links the hazard factors in the event of emergency landing of the designed aircraft, $\mathbb{H}$, with the characteristics of its structure, $\mathbb{S}$, and the emergency landing conditions, $\mathbb{P}$, $\{V\}$ and $\{\Psi\}$. Accordingly, for known arguments: $\mathbb{S}$, $\mathbb{P}$, $\{V\}$ and $\{\Psi\}$, one can find the values of the hazard factors and check the fulfillment of the safety criteria expressed by Eq. (7.4).

In addition to its use at the design stage, this methodology is applicable to aircraft crash investigations, i.e. at the operation stage. An example of such an application was the dynamic analysis of the emergency landing (actually a crash) of the Mi-34 helicopter, carried out at the request of TsAGI (Lyakhovenko et al., 1989; Spirochkin and Schenk, 1990).

The methodology described can serve as a prototype for developing a more sophisticated theoretical approach to the design of new aircraft, including those with non-standard configurations or manufactured using new structural materials. Improvements can also be aimed at analyzing safety conditions in a wider range of emergency landing scenarios and testing more advanced protection systems against hazard factors.

The finite element model required for a refined dynamic analysis that takes into account the details of nonlinear deformation processes and destruction of structural elements of an aircraft when interacting with various landing surfaces obviously should include millions of degrees of freedom. The creation of such a model is possible only at the final phases of design, when the entire set of characteristics of its structure, $\mathbb{S}$, is known with due accuracy. Dynamic analysis of one emergency landing scenario, even using modern high-performance computers, can take hours, and the computing resources required for multivariate simulations are even greater.

The time spent analyzing an emergency landing during aircraft design (or during a crash investigation) is certainly expensive, but it is not as critical as in flight conditions, when an acceptable (from a safety perspective) option for performing an emergency landing must be chosen within a few minutes. Dynamic analysis is feasible at "ground conditions" and at a pace dictated by the available computing equipment, but it is practically inapplicable for forecasting the behavior of the aircraft structure and assessing safety in real time and with the resources available on board the aircraft, i.e. to perform operation 8 in Table 7.2. Thus, the methodology of dynamic analysis is not suitable for direct implementation in CECS, given its application in the event of emergency landing. To support control in critical and emergency situations (emergency landing is just one of many possible accidents) other forecasting methods are needed within the SPARS concept.

7.4.2 Response Surface Method

The above mentioned shortcomings inherent in the methodology of dynamic analysis can be overcome by turning to the method of approximate analysis of the behavior of structures using response surfaces. The response surface is a geometric representation of the response function – the dependence of a variable describing the response of a structure on the arguments characterizing the conditions under which this response occurs. In the case under consideration, a variable describing the response can be any of the hazard factors, and the arguments are $\mathbb{S}$, $\mathbb{P}$, $\{V\}$, $\{\Psi\}$, etc. Thus, we are dealing with a multitude of response functions covered by the generalized function $\mathcal{F}$ and the same multitude of response surfaces. The part of this multitude related to the maximum values of accelerations includes the following functions:

$$\begin{aligned} a_{1x\,max} &= \mathcal{F}_{a1x}(\mathbb{S}, \mathbb{P}, \{V\}, \{\Psi\}, ...); \\ a_{1y\,max} &= \mathcal{F}_{a1y}(\mathbb{S}, \mathbb{P}, \{V\}, \{\Psi\}, ...); \\ a_{1z\,max} &= \mathcal{F}_{a1z}(\mathbb{S}, \mathbb{P}, \{V\}, \{\Psi\}, ...); \\ a_{2x\,max} &= \mathcal{F}_{a2x}(\mathbb{S}, \mathbb{P}, \{V\}, \{\Psi\}, ...); ..., \end{aligned} \tag{7.9}$$

where the subscript numerical index at the maximum value of an acceleration component denotes the observed characteristic point of the structure, equipment or payload, at which this hazard factor can reach a critical level.

A response surface can be visualized only in the case of two arguments, and if there are more, it becomes a hypersurface, which is impossible to depict. A hypersurface can be represented using a set of two- or three-dimensional geometric objects that are its "cross sections" for fixed values of a number of arguments.

The study of the postulated emergency landing of *Buran* (briefly presented in Sect. 7.4.1) showed that, under the accepted scenarios for its implementation, an assessment of safety conditions is possible based on the maximum value of vertical acceleration at the center of gravity of the unit of auxiliary devices. For certain structural characteristics of the aerospace vehicle, $\mathbb{S}$, and rotation angles of its axes, $\{\Psi\}$, the landing conditions, which mainly determine the maximum value of vertical acceleration at this point[7], $a_{y\,max}$, are reduced to the ultimate strength of the landing surface, σ, and the vertical velocity at the moment of its contact, V_y. Thus, we arrive at the response function:

$$a_{y\,max} = \mathcal{F}_{ay}(\sigma, V_y)\,|_{\mathbb{S},\{\Psi\}}, \tag{7.10}$$

and a response surface that allows visualization – see Figure 7.4.

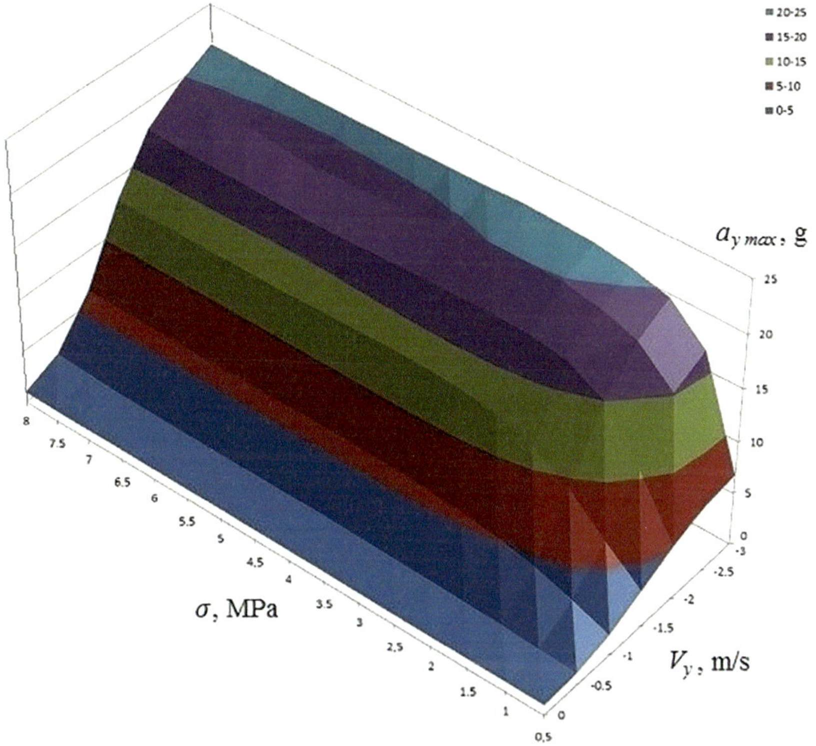

Fig. 7.4 Response surface: dependence of the maximum value of vertical acceleration, $a_{y\,max}$, at the center of gravity of the payload (the unit of auxiliary devices) on the ultimate strength of the landing surface, σ, and the vertical velocity, V_y, in the emergency landing of *Buran*.

This response surface is built on the results of multivariate dynamic analysis in the following ranges of the *Buran* landing conditions: (0.5 MPa $\leq \sigma \leq$ 8 MPa), and (–3 m/s $\leq V_y \leq$ 0). The values of σ and V_y were varied in these ranges, to determine the nodal values of the response function, and then linear interpolation was used to obtain the values at intermediate points. Despite the roughness of this interpolation, the resulting surface clearly reflects the nature of the response function.

[7]The subscript numerical index of the point is omitted here and below.

If necessary, response functions can be determined in terms of other hazard factors and taking into account a larger number of arguments. In the latter case, the response surfaces lose the ability to be presented visually, but the important advantage of the described method remains. This advantage is that the response functions, even when they correspond to multidimensional hypersurfaces, allow one to quickly forecast the behavior of the aircraft structure and estimate the hazard factors during an emergency landing, $\mathbb{H}$, with known characteristics of the structure, $\mathbb{S}$, and expected values of $\mathbb{P}$, $\{V\}$, $\{\Psi\}$ (if these values fall within the studied variation ranges).

The procedure for estimating the hazard factors, $\mathbb{H}$, and checking the fulfillment of the safety criteria (Eq. 7.4), using the known response surfaces do not require large computing resources. The corresponding computational operations, mainly related to interpolation, can be performed by the on-board computer, providing substantiation for the choice of landing surface and kinematic parameters of the aircraft in real time. Tables containing the nodal values of the response functions, and thereby determining the generalized function $\mathcal{F}$, must be generated at the aircraft design stage based on dynamic analysis of a set of expected states of the structure (taking into account, among other aspects, various payloads) and emergency landing scenarios. If necessary, these tables can be adjusted and supplemented at the operation stage – immediately before the flight.

The response surface method is consistent with the principles of critical and emergency control that are fundamental to CECS, including its application to emergency landing. Thus, it appears to be very promising for implementation within the framework of the SPARS concept. However, implementation can be complicated by some response function features – such as non-smoothness or discontinuity, which prevent the determination of response values at intermediate points using certain types of interpolating functions. These features may be related, for example, to the fact that structural characteristics actually vary in the range of possible values not necessarily in a smooth and continuous manner. In some intervals, they, as well as the values of other arguments of the response functions, may be completely absent.

The above difficulties can be overcome by using special mathematical techniques, in particular artificial smoothing and reconstructive filling of lacunae. At the same time, there is an alternative approach that is free from limitations of this kind. It is described in Sect. 7.4.3.

7.4.3 Neural Network Approach

The current level of AI development makes it possible to implement the principles of critical and emergency control envisaged in the SPARS concept, based on neural network technology. The idea of constructing relationships between loading conditions and the response parameters of a structure in its actual state using a neural network was reflected in the regulatory documents of the nuclear industry, in the preparation of which the author took part (SPiR-O-2008; VERLIFE, 2013;

OTT 1.5.2.01.999.0157-2013). This idea, combined with technological advances in recent years, has stimulated further R&D in the field of emergency landing safety, which was interrupted by the closure of the *Energia-Buran* program.

In the first phase of continuing work on this topic, a neural network was developed and an experimental study of its application was conducted using the example of the emergency landing of *Buran*. The development was carried out by engineer Igor V. Uspensky. He also prepared the training data set and processed hazard forecasts for several emergency landing scenarios. The goal of this phase (led by the author), which can be considered a pilot project, was to:

- evaluate the accuracy of forecasting the behavior of the structure and hazard factors in an emergency landing using a neural network;
- estimate the time required for its training and obtaining the forecasts (inferences);
- identify the advantages of the neural network approach compared to dynamic analysis (Sect. 7.4.1) and the response surface method (Sect. 7.4.2).

The neural network, developed taking into account the existing recommendations and experience from preliminary experiments, includes four layers, as shown in Figure 7.5. The first layer, which receives input data, is formed by 80 neurons. The next two, inner layers contain 400 and 30 neurons, respectively. The output layer, from which the generated inferences are taken, is formed by one neuron. The neural network is implemented in the Python programming language using the open source platform TensorFlow (http://tensorflow.org).

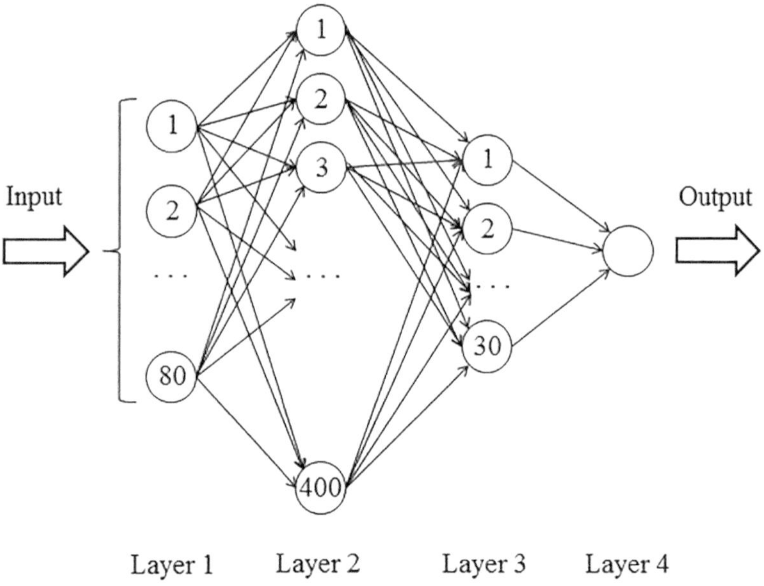

Fig. 7.5 Structure of the neural network developed to forecast the behavior of the structure and hazard factors in an emergency landing.

To perform forecasts, a neural network must be trained on a set of known data characterizing the relationship between input and output. In the situation under consideration, this procedure corresponds to the construction of the generalized function $\mathcal{F}$ in the form of a specific algorithm for obtaining inferences – output variables characterizing the dynamic response of the structure and hazard factors that depend on the input variables, i.e. the arguments of the function.

The pilot project involved the neural network operating in several modes, differing in the number of input variables and the type of data sources. The training set for the main mode covered the variation of two input variables – the ultimate strength of the landing surface, σ, and the vertical velocity of *Buran*, V_y, as well as an output variable – the maximum value of the vertical acceleration at the center of gravity of the unit of auxiliary devices, $a_{y\,max}$. The training was carried out using 475 examples generated on the basis of previously obtained results of the dynamic analysis of the *Buran* emergency landing (this analysis was performed during the design process of the aerospace vehicle and was not part of the described pilot project). The time spent on training, including data exchange with the cloud server hosting the software library of TensorFlow, was no more than 15 s.

The trained neural network performed forecasts of the values of $a_{y\,max}$ for 21 combinations of input values of σ and V_y corresponding to *Buran* emergency landing scenarios that were not included in the training set. The time to obtain the forecasts was about a second. The forecast results are presented in Figure 7.6 against the background of the values found by dynamic analysis (Sect. 7.4.1). As can be seen, the acceleration values generated by the neural network, despite their approximate nature, agree quite well with the results of dynamic analysis, which can be considered more accurate. For the studied emergency landing variants, the differences in the corresponding values do not exceed 4.53%. In a number of cases, the forecasted acceleration values turned out to be lower than those given by dynamic analysis. At this phase of the work, it is not yet clear whether such deviations are random or systematic.

In addition to the main mode, another network operation option was tested. In this option, the number of input variables was increased to three. The third variable was the mass of the aerospace vehicle, which can change according to the transportation of various payloads. The influence of this variable, which is part of the argument $\mathbb{S}$ of the generalized function $\mathcal{F}$, was taken into account by supplementing the training set with the corresponding analytical law. In this neural network operating mode, the feasibility of using data sources other than numerical arrays was assessed. The results confirmed this feasibility.

A neural network similar to the one described can become part of CECS – for example, as software loaded into the on-board computer. The neural network training, i.e. the determination of the generalized function $\mathcal{F}$, should be performed at the pre-operation stages of the aircraft life cycle, including flight tests. The results of dynamic analyses obtained for a variety of assumed states of its structure and emergency landing scenarios are applicable as a training set. If necessary,

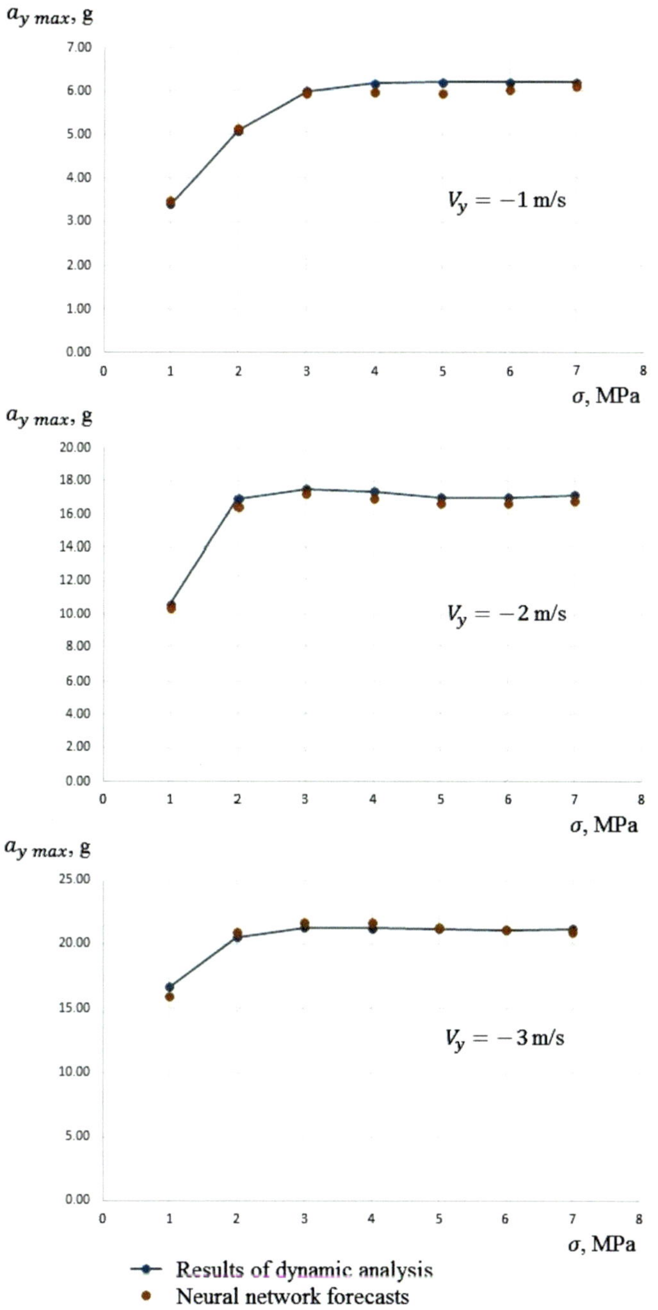

Fig. 7.6 Neural network forecasts in comparison with the results of dynamic analysis for a number of the *Buran* emergency landing scenarios.

additional training can be conducted before transporting cargo (payload) with previously unforeseen characteristics.

The range of training data sources can be expanded to include text, hypertext and graphical data. However, in this case, it is advisable to use a composite neural network that includes several branches, each of which specializes in a certain type of information. Then the generalized forecast result obtained in the output layer, where the branches are connected, will have a higher accuracy.

In addition to forecasting the behavior of the structure and hazard factors during an emergency landing, the neural network approach is obviously applicable to other operations listed in Table 7.2 that consume significant computer resources, as well as to control of other critical and emergency situations. Such opportunities should be substantiated in the course of further R&D. The theoretical details of neural network technology, omitted above, are presented in Sect. 8.1.2. They help assess the overall potential of this technology, including its application to minimizing uncertainty in design. Some possible applications are described in Sects. 8.2.2, 8.3.6 and 8.4.

7.5 Bionic Aspects of the SPARS Concept as Applied to Emergency Landing

7.5.1 Embodiment of Behavioral Bionic Principles

The bionic ideas incorporated in the SPARS concept, which meet the required properties of smartness and proactivity (Sect. 6.4.1) and belong to its innovative aspects, are implemented according to the principles of behavioral bionics (Sect. 6.4.3). These principles, which also include an active aspect (response to immediate impact), involve replicating the behavior of a living being when faced with an adverse event that poses a threat to existence. For aircraft, such replication can cover the following functions:

- observation of the environment and the state of aircraft, which allows identifying changes occurring during the flight, taking into account their potential danger;
- preventive response to changes that may presumably lead to deterioration of normal flight conditions (proactive control);
- detection of deviations from normal flight conditions, assessment of the degree of their danger in real time and response to avoid or mitigate expected harm (reactive control);
- protection from hazard factors in the event of an uncontrolled development of an emergency situation (its transformation into an accident);
- recovery (at least partial) after an accident;
- learning from the emergency situation or accident, to deal more effectively with similar events in the future.

These functions are performed by CECS. The work of this system in the event of emergency landing is described in Sect. 7.3.4. The automatic operations listed in Table 7.2 should replace the pilot's actions in a situation where possible negative manifestations of human factors – such as erroneous assessments and decisions, delayed reaction, stress or inaction – can lead to fatal consequences. As shown in Sect. 7.4.3, hazard assessment operations within the context of critical and emergency control can be performed efficiently and quickly using a neural network. According to current understanding, this AI element works similarly to the brain of a living being, therefore, its application also reflects the behavioral bionic principles inherent in the SPARS concept. In the case of using a neural network, data from each performed emergency landing must be stored (operation 16 in Table 7.2) to expand the training set. This operation corresponds to the aforementioned learning function (item 6), the purpose of which is to apply the acquired knowledge to similar situations in the future. This also falls within the framework of behavioral bionics.

7.5.2 On the Applicability of Bionic Ideas to Aircraft Recovery after Emergency Landing

The possibility to recover an aircraft after emergency landing depends on the extent of structural damage, the adequacy of the damage assessment, and the recovery technologies used. The extent of structural damage is determined by the "softness" of the emergency landing in combination with the construction principles aimed at minimizing and localizing damage. Providing "soft" emergency landing conditions in an evolving situation is part of the functions of CECS, which embodies, as indicated above, the behavioral bionic principles. Similar principles are implemented when this system carries out damage assessment – through post-accident analysis of the aircraft state in accordance with operation 15 in Table 7.2.

Traditional technologies used to recover an aircraft after emergency landing (when recovery is considered possible) rely on repairing or replacing damaged structural elements and systems. The ability to repair or replace damaged aircraft parts is linked to bionic principles not at the level of functions, but through the overall goal of ensuring a long life cycle.

Along with these technologies, available in production facilities, bionic ideas are being developed in the frame of the SPARS concept aimed at the self-recovery of damaged aircraft elements at the site of an emergency landing. These ideas are, to a certain extent, consistent with the principles of resilient engineering incorporated into this concept. They follow the direction of self-healing materials and structures (Ghosh, 2009; Speck et al., 2013; Atay, 2022; Crawford, 2022), but, compared to current advances, they are focused on self-healing from more extensive damage, extending beyond fatigue cracks in metal alloys and plastics or delamination in composite materials. The development and application of metamaterials with the appropriate properties appears to be the most promising way to achieve this goal.

References

AC 25-17A (2016) Advisory circular (with Change 1). Subject: Transport airplane cabin interiors. Crashworthiness handbook (Date of Change 1: 05/24/16). Federal Aviation Administration. https://www.faa.gov/documentLibrary/media/Advisory_Circular/AC_25-17A_CHG-1.pdf. Accessed September 24, 2025.

AC 29-2C (2023) Advisory circular (with Changes 1-9). Subject: Certification of transport category rotorcraft (Date of Change 9: 6/23/23). Federal Aviation Administration. https://www.faa.gov/documentLibrary/media/Advisory_Circular/AC_29-2C_CHG1-9.pdf. Accessed September 24, 2025.

AP-25 (2009) Aviatsionnye Pravila. Chast 25. Normy lyotnoi godnosti samolyotov transportnoi kategorii (*Aviation rules. Part 25. Airworthiness standards for transport category airplanes*), 3rd edn. Interstate Aviation Committee, Moscow (in Russian).

ASN (2009) Aviation Safety Network – ASN Aviation Safety WikiBase. https://asn.flightsafety.org/wikibase/321581. Accessed September 22, 2025.

ASN (2019) Aviation Safety Network – ASN Aviation Safety WikiBase. https://asn.flightsafety.org/asndb/319362. Accessed September 22, 2025.

ASN (2025a) Aviation Safety Network – ASN Aviation Safety WikiBase. https://asn.flightsafety.org/wikibase/468275. Accessed September 22, 2025.

ASN (2025b) Aviation Safety Network – ASN Aviation Safety WikiBase. https://asn.flightsafety.org/wikibase/468716. Accessed September 22, 2025.

Atay, H.Y. (2022) Self-healing composite materials. In Inamuddin, T. Altalhi, S.M. Adnan (eds.) *Aerospace polymeric materials*. Chapter 6. Scrivener Publishing LLC, Beverly, MA. pp. 137–154.

CFR Part 25. Code of Federal Regulations/Title 14/Chapter I/Subchapter C/Part 25—Airworthiness Standards: Transport Category Airplanes. https://www.ecfr.gov/current/title-14/chapter-I/subchapter-C/part-25. Accessed September 22, 2025.

Crawford, M. (2022) 7 Self-healing materials expand the limits of engineering design. ASME, November 22, 2022 (Updated: October 24, 2024). https://www.asme.org/topics-resources/content/7-self-healing-materials-expand-the-limits-of-engineering-design#:~:text=Self%2Dhealing%20materials%2C%20when%20damaged,visible%20or%20even%20measurable%20problems. Accessed October 05, 2025.

Cronin, R. (2025) How common is an emergency landing after pilot 'collapses' on EasyJet flight. *Daily Record*, February 10, 2025. https://www.dailyrecord.co.uk/lifestyle/how-common-emergency-landing-after-34649677. Accessed January 17, 2026.

CS-25 (2018) Certification specifications for large aeroplanes (Amendment 21). European Aviation Safety Agency, Brussels.

ENLG (1985) Edinye Normy Lyotnoi Godnosti grazhdanskikh transportnykh samolyotov stran–chlenov SEV (*Uniform airworthiness standards for civil transport airplanes of the Comecon member states*). TsAGI, Moscow (in Russian).

FAA-H-8083-3C (2021) *Airplane Flying Handbook*. U.S. Department of Transportation, Federal Aviation Administration.

FAA-H-8083-6 (2009) *Advanced Avionics Handbook*. U.S. Department of Transportation, Federal Aviation Administration. Updated May 2, 2013.

Frings, G. (1998) FAA Crashworthiness Research Program. *International Aircraft Fire and Cabin Safety Research Conference*. Atlantic City, NJ, 17 November 1998.

Ghosh, S.K. (ed.) (2009) *Self-healing materials: Fundamentals, design strategies, and applications*. Wiley–VCH Verlag GmbH & Co. KGaA, Weinheim, Germany.

Guida, M., Marulo, F. and Abrate, S. (2018) Advances in crash dynamics for aircraft safety. *Progress in Aerospace Sciences*, 98:106–123.

IAEA (2016) IAEA Safety Standards Series No. SSR-2/1 (Rev.1). Safety of nuclear power plants: Design. Specific safety requirements. International Atomic Energy Agency, Vienna.

IdeaConnection (2015) Civil aviation safety systems. *IdeaConnection by Planbox*, 17 May 2015. https://www.ideaconnection.com/new-inventions/civil-aviation-safety-systems-09615.html. Accessed September 30, 2025.

Jackson, K.E. and Fasanella, E.L. (2003) A survey of research performed at NASA Langley Research Center's Impact Dynamics Research Facility. American Institute of Aeronautics and Astronautics, Reston, VA; ResearchGate, June 2003. https://www.researchgate.net/publication/266370987. Accessed September 24, 2025.

Jackson, K.E., Boitnott, R.L., Fasanella, E.L., Jones, L.E. and Lyle, K.H. (2006) A summary of DOD-sponsored research performed at NASA Langley's Impact Dynamics Research Facility. *Journal of American Helicopter Society*, 51(1):59–69.

Joseph, C. (2012) Attention first-class passengers, this is your captain speaking... Crash test proves that it is much safer to sit in the back of the plane. *MailOnline*, 16 September 2012. https://www.dailymail.co.uk/news/article-2203888/The-ultimate-crash-test-dummy-What-passengers-didnt-want-hear-safer-sit-back.html. Accessed September 22, 2025.

Lau, S.T., Said, M.R. and Yaakob, M.Y. (2012) On the effect of geometrical designs and failure modes in composite axial crushing: a literature review. *Composite Structures*, 94(3):803–812.

Liang, X., Fu, H. and Crosby, A.J. (2022) Phase-transforming metamaterial with magnetic interactions. *Proceedings of the National Academy of Sciences*, 119(1):1–8.

Lyakhovenko, I.A., Shenk, Y.V. and Spirochkin, Y.K. (1989) On the possibility of numerical simulation of a helicopter emergency landing. Scientific and technical report NTO 03-5942. TsAGI, Zhukovsky, Moscow region (in Russian).

Meuleau, N., Plaunt, C., Smith, D.E. and Smith, T. (2009a) A comparison of risk sensitive path planning methods for aircraft emergency landing. *ICAPS-09 Workshop on Bridging the Gap Between Task and Motion Planning* (BTAMP), Thessaloniki, Greece.

Meuleau, N., Plaunt, C., Smith, D.E. and Smith, T. (2009b) An emergency landing planner for damaged aircraft. *Proceedings of the Twenty First Innovative Applications of Artificial Intelligence Conference*. (IAAI-09), Pasadena, California. AAAI Press.

NASA (2002) Intelligent flight control system. NASA Facts, FS-2002-09-076-DFRC, National Aeronautics and Space Administration, Dryden Flight Research Center, Edwards, CA. https://www.nasa.gov/wp-content/uploads/2021/09/120321main_FS-076-DFRC.pdf. Accessed September 29, 2025.

NP-001-15 (2015) Federalnye normy i pravila v oblasti ispolzovaniya atomnoi energii. Obshchie polozheniya obespecheniya bezopasnosti atomnykh stantsii (Federal regulations and rules in the field of atomic energy use. General provisions for ensuring safety of nuclear power plants). Federal Environmental, Technological and Nuclear Supervision Service of Russia, Moscow (in Russian).

OTT 1.5.2.01.999.0157-2013 (2013) Opornye konstruktsii elementov atomnykh stantsii s vodo-vodyanymi energeticheskini reaktorami. Obshchie Tekhnicheskie Trebovaniya (Support structures for elements of nuclear power plants with water-water power reactors. General technical requirements). JSC Concern Rosenergoatom, Moscow (Contributors: Spirochkin Y.K., Obushev A.E., Filimonov S.V., Sorokin A.N., Tarakanov P.V. et al.). https://meganorm.ru/Index2/1/4293736/4293736524.htm. Accessed October 04, 2025 (in Russian).

Patrushev, V.S., Pletnyov, I.V. and Spirochkin, Y.K. (1989) Aviatsionno-kosmicheskii trenazhyor (Air-spacecraft simulator). Patent SU 1799479, November 24, 1989 (in Russian)

Rossier, R.N. (2011) Differentiating precautionary landings, forced landings, and ditching. Flight Training Magazine: Emergency Landings. FAA Safety Team. https://www.faasafety.gov/files/events/EA/EA23/2011/EA2336098/FlightEmergencies.pdf. Accessed September 22, 2025.

Sharma, S. (2024) Plane makes emergency landing without landing gear after circling for 4 hours in Newcastle airport drama. *Independent*, 13 May 2024. https://www.independent.co.uk/travel/news-and-advice/newcastle-emergency-landing-plane-airport-b2543941.html. Accessed September 22, 2025

SPiR-O-2008 (2009) Svod Pravil i Rukovodstv po Opornym konstruktsiyam elementov AES s WWER (Code of rules and guides on supports for elements of nuclear power plants with WWER). Standard of organization. ENES, Moscow (Contributors: Evropin S.V., Filatov V.M., Golovlyov Y.N., Obushev A.E., Rodchenkov B.S., Spirochkin Y.K. et al.) (in Russian).

Speck, O., Schlechtendahl, M., Borm, F., Kampowski, T. and Speck, T. (2013) Bio-inspired self-healing materials. In P. Fratzl, J.W. Dunlop, R. Weinkamer (eds.) *Materials design inspired by nature: Function through inner architecture* (Series: RSC Smart Materials), Volume 4. The Royal Society of Chemistry, Cambridge, UK. pp. 359–389.

Spirochkin, Y.K. (1987) Metodika raschyota dinamicheskogo nagruzheniya gruzhov pri avariinoi posadke letatelnogo apparata (Method for analyzing dynamic loads on items of mass during an aircraft emergency landing). Dissertation. NPO *Energia*, Kaliningrad (now Korolyov), Moscow region (in Russian).

Spirochkin, Y.K. (1988) Iskhodnye dannye na provedenie eksperimentalnykh issledovanii parametrov reaktsii posadochnoi poverkhnosti s ispolzovaniem zhyostkikh maketov. Otchet P 26478-017 (Baseline data for experimental studies on the landing surface response parameters using rigid mockups. Report P 26478-017). NPO *Energia*, Kaliningrad (now Korolyov), Moscow region (in Russian).

Spirochkin, Y.K. (1993) Computer simulation of structural dynamics in accidents. *Mathematical Models and Computer Simulations*, 5(6):85–103 (in Russian)

Spirochkin, Y.K. (2019) Bezopasnost rossiiskikh AES s tochki zreniya inzhenera-mekhanika (*Safety of Russian NPPs from the viewpoint of mechanical engineer*). SUPER Izdatelstvo, Saint Petersburg (in Russian).

Spirochkin, Y. (2023) *Human factors and design*. Springer, Singapore

Spirochkin, Y.K. and Shenk, Y.V. (1990) Chislennoe modelirovanie avariinoi posadki vertoleta (Numerical simulation of a helicopter emergency landing). Problemy Udara, Razrusheniya i Tekhnologii *(Problems of Impact, Destruction and Technology)*, 10(4):62–63 (Publishing house VNITsUR-Resurs, Riga) (in Russian).

Stengel, R.F. (1992) Intelligent flight control systems. *IMA Conference on Aerospace Vehicle Dynamics and Control,* Cranfield Institute of Technology, Bedford, UK, September 7-10, pp. 151–172.

Synbiosys (2021) Phase change armour. Synbiosys. Material science. https://www.synbiosys.co/material-science. Accessed September 30, 2025.

VERLIFE (2013) Guidelines for integrity and lifetime assessment of components and piping in WWER nuclear power plants. IAEA, Vienna. Appendix F – Component and piping supports (Contributors: Obushev A., Spirochkin Y.).

8 The Potential of Artificial Intelligence in Minimizing Uncertainty

This chapter explores the theme of using AI to minimize uncertainty in design, which is highly relevant but under-researched. An overview of currently available AI elements, conducted from an engineering design perspective, shows that neural network technology is the dominant one. This technology is applicable, in particular, to such problems as recognizing hidden structural anomalies detected during in-service monitoring and diagnostics, or controlling aircraft in critical situations created by humans in flight. An engineering formulation of these problems and a solution methodology capable of reducing uncertainty are presented. The potential of neural network technology for solving some other problems is revealed. While the use of AI elements in design tasks promises significant advances, when applied to safety-critical systems, it may create new risks. Opportunities for mitigating the risks associated with AI implementation are discussed at the end of the chapter.

8.1 Artificial Intelligence from an Engineering Design Perspective

8.1.1 General Overview of AI Development and Its Elements

The term "artificial intelligence", or AI, has appeared several times in this book. In Sect. 4.3.4 it was mentioned that the creation of AI in the form of neural networks can be viewed as a reconstruction of the human brain. In Sects. 6.4.1 and 6.4.3 the application of AI within the SPARS design concept was noted. Section 7.4.3 illustrated the capabilities of neural networks using the example of ensuring safety in an aircraft emergency landing. This example demonstrated the fundamental advantages of neural network technology for control in critical and emergency situations. One of the advantages is the high speed of data processing, which makes it possible to outpace the dangerous development of such situations and ensure the controllability of the technical system, taking into account its current state, which may differ from that envisaged in the design. Another advantage is the training of a control system with an embedded neural network based on heterogeneous information, the nature of which complicates the application of classical mathematical methods. It is also worth mentioning the ability to incorporate similar new data obtained in real time into the training process, which in principle allows the inferences to be refined as the situation evolves.

These advantages seem to be highly attractive for using AI to solve other problems that may arise in the post-project stages of the life cycle of modern technical

DOI: 10.1201/9781003670575-8

systems. Therefore, exploring the potential of AI that designers could leverage in addressing such problems is extremely relevant. AI is a rapidly developing field of technology, currently receiving colossal financial and intellectual investment. Unsurprisingly, such development is accompanied by significant information noise and the speculative nature of some activities in this field. At the same time, there is growing concern in society, including scientists and technical specialists, about the impact of AI on the future, given that not all directions and limits of AI evolution are clear and effectively regulated today (Copeland, 2025). In these circumstances, it is important to understand what AI objectively is and how this technology can be effectively implemented in engineering design.

Since John McCarthy, an American computer programming specialist, coined the term "artificial intelligence" in 1956, it has become associated with computer systems. This term has the following meanings: 1) the property of computer systems to perform functions inherent in human intelligence, and 2) the field of science and technology associated with the creation of such systems. The first part of the term, "artificial," is understood unambiguously, but the second, "intelligence," can be misleading, since it can be interpreted to mean that computer systems at a certain level of development are capable of possessing consciousness and thinking. The question of whether they are capable of thought was first posed by Alan Turing, an English mathematician and one of the founders of computing. However, even today, more than 75 years after his seminal work (Turing, 1950), there is no exact criterion the fulfillment of which would indicate that a computer has intelligence. The main obstacle to developing such a criterion is the lack of understanding of what intelligence is, as well as reason, thought, or consciousness. Therefore, the term "artificial intelligence", in which only the first word (an adjective) is justified, while the second word (a noun) conceals something vague, is not entirely appropriate. Nevertheless, the term has caught on and become widely used.

Currently, there are three distinct areas of R&D in the field of AI (Copeland, 2025):

- artificial general intelligence (AGI);
- applied AI;
- cognitive simulation.

The ultimate goal of efforts in the first area is to create a machine whose general intellectual ability would be indistinguishable from that of a human. This goal has not yet been achieved, but one of the impressive results is large language models, LLMs, – see, for example, Minaee et al. (2024); Annepaka and Pakray (2025). Applied AI is considered as a development of *advanced information processing*, a direction that has existed for decades, and this development is aimed at creating "smart" computerized tools suitable for effectively solving various practical problems. Today, the range of such tools covers expert systems, including those implementing medical and technical diagnostics, classification and inference systems, pattern recognition systems, forecasting systems, etc. (Enríquez et al., 2019; Yakimov and Maltsev, 2022; Yang and Zhu, 2024; Neupane et al.,

2025). Cognitive simulation is a research approach in neuroscience and cognitive psychology, used mainly to test theories about the workings of the human mind (Roth et al., 1992; Frischkorn and Schubert, 2018; Zhao and Smidts, 2021). Thus, from the point of view of the application of AI in the design of technical systems, as well as equipping the designed systems with AI elements, the second area—applied AI – is of greatest interest.

The signs of "intelligence" of computer programs that implement advanced information processing compared to "conventional" software (for example, those performing engineering calculations) are considered to be the abilities to:

- find stable patterns and statistically significant relationships in large arrays of data characterizing the object of analysis and often representing heterogeneous information flows;
- use fuzzy logic capable of handling the "small-large", "better-worse" and other similar poorly quantifiable relationships;
- self-learn by automatically changing the data processing models (algorithms), embedded in the programs, in accordance with the data being processed (primarily with that part of it that serves as a training set).

Since all these manifestations of "intelligence" are realized by electronic circuits that perform arithmetic and logical operations, AI (at least for now) does not go beyond the scope of information technology, being its specific, rapidly developing branch. Modern aviation and aerospace systems, NPPs, processing facilities and other safety-critical and expensive technical systems rely heavily on computerized information support for their life cycle (Sect. 6.2.2). The purpose of implementing AI in these systems is to give this support a new quality – the plasticity inherent in human intelligence, while eliminating the negative manifestations of human factors.

R&D in the field of AI is conducted in two main, somewhat competing directions:

- symbolic artificial intelligence (or logic-based AI) and
- artificial intelligence based on artificial neural networks (which imitate the structure and functioning of the brain).

Symbolic AI models knowledge through logical structures and uses tools such as logic programming, production rules, semantic nets and frames (Van Harmelen et al., 2008; Garnelo and Shanahan, 2019; Thomason, 2024). The results of the work in this direction are expert systems, symbolic mathematics, automated theorem provers, ontologies, the semantic web, and automated planning and scheduling systems. The symbolic AI approach dominated until approximately the mid-1990s.

Currently, the second approach to creating AI is the most widespread. A distinctive feature of a neural network is that it stores memory of how to handle information containing certain signs in a network structure consisting of interconnected artificial neurons arranged in layers, as shown, for example, in Figure 7.5. The signs in question are typically not explicitly defined, but become known to the network during the training process and must then be recognized when it is used in various contexts. More details about neural network technology are presented below, in Sect. 8.1.2.

Each of these approaches has its advantages and shortcomings. Using the former while compensating for the latter appears possible within the framework of a hybrid approach to AI development that has recently emerged (Ying et al., 2025).

8.1.2 Some Details of Neural Network Technology

Let us take a closer look at neural network technology, the most widely used embodiment of AI today. Problems solved by neural networks (we mean artificial neural networks, but we omit the adjective "artificial") typically fall into the following categories:

1) classification of data, i.e. assigning them to certain classes;
2) recognition of observed objects;
3) forecasting, i.e. producing inferences about future developments based on given information relating to the past;
4) autonomous control of objects.

The work of a neural network in solving the listed problems can be considered as interpolation using a certain function that approximates the dependence of the desired output variable (numerical value or other inference characteristic), Y, on input data, X_1, X_2, ... X_d, (d is the number of arguments of the function, i.e., dimensionalities of the domain of definition of Y or its subdomains, considered as arguments). The specificity of this function, which fits for multidimensional dependencies, lies in its construction as a superposition of functions of one variable (Kolmogorov, 1956, 1957; Arnold, 1958; Lorentz, 1962; Lai and Shen, 2025):

$$Y = Y(X_1, X_2, \dots X_d) = \sum_{j=0}^{j=2d} g\left(\sum_{i=1}^{i=d} \psi_j(X_i)\right), \tag{8.1}$$

where g is the so-called outer function, and ψ_j are inner functions ($j = 0, 1, \dots 2d$).

In modern versions of neural network technology, interpolation is performed recursively using the following algorithm:

$$\begin{aligned} \{x^{(0)}\} &= \{X\} = [X_1 \;\; X_2 \dots X_d]^{\mathrm{T}}; \\ \{x^{(l)}\} &= \vartheta\left([w^{(l)}]\{x^{(l-1)}\} + \{b^{(l)}\}\right); l = 1, 2, \dots, L-1; \\ Y &= [w^{(L)}]\{x^{(L-1)}\} + \{b^{(L)}\}, \end{aligned} \tag{8.2}$$

where ϑ is an activation function;

$[w^{(l)}]$ is the matrix of weighting coefficients for the l-th layer of neurons, $[w^{(l)}] = [w_1^{(l)} \;\; w_2^{(l)} \;\; \dots \;\; w_{N_l}^{(l)}]$ (N_l designates the number of neurons in this layer);

$\{b^{(l)}\}$ is the bias vector of the l-th layer, $\{b^{(l)}\} = [b_1^{(l)} \;\; b_2^{(l)} \;\; \dots \;\; b_{N_l}^{(l)}]^{\mathrm{T}}$;

L is the number of layers.

The input data, $\{X\}$, is fed into the input layer, forming the vector $\{x^{(0)}\}$. The value of each component of the vector $\{x^{(l)}\}$ in each neuron of the internal l-th layer is obtained by applying the activation function, ϑ, to the weighted sum of the values that came to the given neuron from the neurons of the previous layer, $l - 1$, taking into account the bias vector $\{b^{(l)}\}$. The activation function performs a nonlinear transformation of the components of the output vector of the previous layer and normalizes them on a scale from 0 to 1 (see, for example, https://developers.google.com/machine-learning/crash-course/neural-networks/activation-functions). The components of the bias vector act as constant input values for neurons in the subsequent layer, allowing the activation function to be shifted left or right along the argument axis. Bias is an option in software libraries used to construct neural networks. It can improve the neural network's ability to match the input data. The output variable Y is generated in the output layer (L).

The weighting coefficients are determined automatically during neural network training on known combinations of $\{X\}$ and Y. The input data can be any variables that influence the desired output, as well as any signs that characterize relationships between the input and output, observed in the subdomains into which the domain of definition of the function Y can be divided, i.e. the ranges of the variation of its arguments.

Methods for adjusting the weighting coefficients of a neural network to minimize errors in the interpolation results during training are the subject of the discipline known as *machine learning* (Goodfellow et al., 2016). The procedure for determining these coefficients using known combinations of input and output is usually called *supervised learning*. The most common method for training neural networks today is *backpropagation* (Nielsen, 2015). In fact, it is the weighting coefficients that carry the memory of how the trained neural network should respond to input data.

In some cases, neural networks are trained in the process of *unsupervised learning*, where the network is given a set of input data without any associated output and is tasked with detecting patterns, without a precise specification of what to look for. Such an approach can be used in data mining (Zwass, 2025).

Unlike classical interpolation methods, due to the complexity of the procedure for determining weighting coefficients, for neural networks used in practice, the exact form of the approximating function obtained as a result of machine learning cannot be presented explicitly. Therefore, it can be difficult to understand how the network arrives at its inference. In this respect, a neural network is like a black box whose operation is impossible to verify using conventional mathematical and engineering methods. It is this opacity that is the main theme of criticism of neural network technology.

However, despite these somewhat discouraging circumstances, and precisely because of the use of specific approximating functions with many parameters that can be adjusted to a given problem, neural networks provide a more flexible and powerful tool for analyzing various datasets, including images. The favorable properties stem from the ability of neural networks to represent very complex nonlinear relationships between data, among them multidimensional ones, without being limited to a predefined function type. Furthermore, it is worth noting their ability to detect patterns that are not immediately obvious.

8.2 Advanced Diagnostics (Recognition) of Hidden Structural Anomalies

8.2.1 Diagnostic Models

To detect hidden anomalies in technical systems that may manifest themselves negatively in the post-design stages of the life cycle and lead to deviations from design operating modes, diagnostic models are used. Advanced diagnostic models also make it possible to determine quantitative data of hidden anomalies in order to draw conclusions about their impact on safety and performance. The concept of a diagnostic model was introduced in Sect. 5.5.1 and was followed by the remark that for complex systems consisting of many elements and characterized by a variety of working and accompanying processes, such a model can be a nontrivial mathematical object. Below this concept is explained using the example of anomalies in the characteristics of the system structure. Limiting attention to structural anomalies does not undermine the generality of the methodology, but rather provides the necessary concreteness.

Detection of hidden structural anomalies is based on the physical relationship between deviations in the characteristics of a structure, $\delta\mathbb{S}$, relative to the nominal (design) state and signs of such anomalies that are observed in the mechanical response of this structure to operational or test loads, and can be considered deviations from the expected behavior corresponding to the nominal state, $\delta\mathbb{R}$:

$$\delta\mathbb{R} = \mathcal{F}(\delta\mathbb{S})|_{\{F\}}, \tag{8.3}$$

where the generalized function $\mathcal{F}$ mathematically expresses the mentioned relationship between two sets: $\mathbb{S}$ and $\mathbb{R}$ when applying a load described by the column vector $\{F\}$.

For example, if a detachable, fixed mechanical connection between two parts of a structure loosens or collapses, causing one part to move relative to the other, this anomaly can manifest itself in the form of new vibration modes in the spectrum of the dynamic response of the structure, which are absent in its nominal state, and (or) changes in the vibration displacement amplitudes. These new vibration modes and changes in vibration amplitudes serve as diagnostic signs of the mentioned anomaly. However, some diagnostic signs may appear under certain loading conditions but be absent under others. Therefore, not all hidden anomalies, at least until they reach a sufficient level characteristic of known defect types (i.e., while they remain latent defects), are detectable under operational loads, and their detection may require testing under special loading conditions. This circumstance determines the importance of using the active vibration diagnostics method (Sect. 6.3).

A diagnostic model contains a relationship that is inverse to that described by Eq. (8.3): with the help of this model, deviations from the nominal state, or nominal values of the characteristics of the structure, are expressed through the observed diagnostic signs, $\delta\mathbb{R}$:

$$\delta\mathbb{S} = \mathcal{F}^{-1}(\delta\mathbb{R})|_{\{F\}}. \quad (8.4)$$

The form of the diagnostic model can be qualitative or quantitative depending on the nature of the diagnostic sign it is intended to analyze. For example, in the above case (loosening or collapse of a detachable mechanical connection), when using vibration diagnostics in its simplest form – without special recording equipment – the diagnostic sign indicating the unacceptable gap between contacting surfaces of parts and their relative movement is a specific noise (periodic knocking) due to the collision of these parts at a certain frequency. The qualitative nature of the deviation in the characteristics of the structure (the emergence of a defect in the connection) and the diagnostic sign (specific noise), taking into account the cause-and-effect relationship between them, determines the possibility of using a qualitative diagnostic model obtained by logically inverting this relationship – e.g., in the form given in Table 8.1.

Table 8.1 An example of a diagnostic model in qualitative form.

Monitored characteristic of the structure	*Loading conditions*	*Observed diagnostic sign*	*Diagnostic conclusion*
State of detachable mechanical connection	Vibration	A noise in the form of a periodic knocking sound	Defect in connection (loosening or collapse)

Diagnostic models in qualitative form can be multifactorial, i.e., they can correspond to many monitored characteristics of the system as factors of its state (structural health and condition) and describe a number of diagnostic signs. Some of the diagnostic signs can be associated with several possible anomalies, and detection of a specific anomaly can only be achieved if additional conditions are met. For example, if a technical system contains moving mechanical connections (keyways, gears, etc.), a specific noise during their work can indicate wear on the contact surfaces of the connection parts, leading to increased gaps between them and multiple impacts. The presence of such connections, their location, and the nature of the noise can all serve as additional conditions that direct the diagnostic process along the right path.

A complex diagnostic model capable of analyzing multiple diagnostic signs and checking a row of conditions can be implemented using symbolic AI, e.g., in the form of an expert system. Expert diagnostic systems are primarily focused on detecting anomalies in a system and recognizing them by assigning them to a specific defect category. They do not always provide the possibility to determine the quantitative parameters of anomalies, i.e., to recognize them in the true sense of the word.

If the observed diagnostic signs can be expressed quantitatively, and the goal of recognition is the quantification of anomalies, then diagnostic models in quantitative form should be used. As an example, let us consider a pipeline transporting a liquid medium that has a corrosive and (or) erosive effect on the inner surface of the pipe. This effect leads to a decrease in the pipe wall thickness (its thinning) over time. Such an effect is taken into account in the relevant provisions of regulatory documents used during design – see, in particular, PNAE G-7-002-86. However, these provisions specify average, conservative values for the rate of thinning. In the framework of life cycle management based on actual state, the actual values of wall thickness decrease in various pipeline zones are of interest. These values can be estimated based on in-service monitoring and diagnostic data. Below are the principles for constructing a quantitative diagnostic model that is applicable in combination with the active vibration diagnostics method (Sect. 6.3).

This method provides for the experimental determination of the actual natural frequencies of pipeline vibrations (we will not consider the possibility of recording vibration mode shapes for now). Comparing the actual frequencies with the corresponding nominal values yields a column vector of frequency deviations, which serve as diagnostic signs of the presence of structural anomalies in the pipeline:

$$\delta\mathbb{R} = \{\delta f\} = [\delta f_1 \quad \delta f_2 \quad \ldots \quad \delta f_m]^{\mathrm{T}}, \tag{8.5}$$

where m is the number of analyzed natural frequencies and, accordingly, diagnostic sings.

The characteristics of hidden structural anomalies to be recognized (identified) are the deviations in the values of the pipeline wall thickness from the nominal parameters in a number of its zones:

$$\delta\mathbb{S} = \{\delta h\} = [\delta h_1 \quad \delta h_2 \quad \ldots \quad \delta h_n]^{\mathrm{T}}, \tag{8.6}$$

where n is the number of pipeline zones to be analyzed.

In this case, the diagnostic model, in principle, can be obtained by inverting the generalized function $\mathcal{F}$ as presented in Eq. (8.4). For the inversion operation to be performed, the following conditions must be met:

- the analytical functions $\mathcal{F}_{ij}$, which constitute the generalized function $\mathcal{F}$ and establish dependencies between δf_i and δh_j ($\delta f_i = \mathcal{F}_{ij}(\delta h_j)$; $i = 1, 2, \ldots, m;$ $j = 1, 2, \ldots, n$), are continuous and differentiable;
- the deviations of the frequencies, δf_i, and the values of wall thickness, δh_j, are small (which ensures the uniqueness of the functions $\mathcal{F}_{ij}$);
- the number of diagnostic signs is equal to the number of pipeline zones to be analyzed, i.e., $m = n$;
- the determinant containing partial derivatives is not equal to zero:

$$\left|\frac{\partial(\delta f_i)}{\partial(\delta h_j)}\right| \neq 0 \;\; (i, j = 1, 2, \ldots, n). \tag{8.7}$$

If the analytical functions $\mathcal{F}_{ij}$ are linear, then the diagnostic model corresponding to Eq. (8.4) takes on a simple form:

$$\{\delta h\} = [A]^{-1}\{\delta f\}, \tag{8.8}$$

where $[A]$ is the matrix containing the sensitivity coefficients of diagnostic signs – deviations in the natural frequencies of the pipeline vibrations – to characteristics of hidden structural anomalies – deviations in the values of the wall thickness in various pipeline zones:

$$[A] = [A_{ij}] = \left[\frac{\delta f_i}{\delta h_j}\right] \quad (i, j = 1, 2, ..., n). \tag{8.9}$$

Sensitivity coefficients A_{ij} can be determined as finite differences based on data from relevant laboratory experiments or calculations. Thus, the diagnostic model can be constructed as a computational algorithm implementing linear algebra operations with the generated matrix $[A]$ and the column vector $\{\delta f\}$ obtained using the monitoring data. The result of diagnostics and recognition is the column vector $\{\delta h\}$ containing the quantitative characteristics of the structural anomalies in the form of a decrease in the wall thickness values in the considered pipeline zones, averaged over the length or area of each zone.

In many practical cases, the recognition problem may be more complex. For example, the precise location, shape, and size of local anomalies are of interest, which leads to a significant increase in the dimensions of the sets $\delta\mathbb{R}$ and $\delta\mathbb{S}$. In addition, the above conditions for the functions $\mathcal{F}_{ij}$ and the elements of the sets $\delta\mathbb{R}$ and $\delta\mathbb{S}$ may not be met. In such cases, the quantitative diagnostic model can be constructed as a neural network. The application of neural network technology for recognizing structural anomalies within the framework of the active vibration diagnostics method is described below, in Sect. 8.2.2.

8.2.2 Application of Neural Network Technology in the Active Vibration Diagnostics Method

The active vibration diagnostics method described in Sect. 6.3 enables the detection of hidden structural anomalies, including defects in the form of macrocracks, based on distortions in the vibration displacement field recorded at resonant frequencies using holographic interferometry. This method is applicable both for factory quality control at the manufacturing stage of technical system elements, particularly pipes and pressure vessels, and for in-service monitoring and diagnostics. Its current implementation requires human intervention in analyzing the interference pattern representing the distribution of vibration displacement amplitudes at each local resonance. During the development and testing of this method, the possibilities of automating such analysis remained unexplored. Regarding the further development of the method, its potential for recognizing the type, shape and size of hidden anomalies was noted, including the use of automated procedures (Sect. 6.3.5).

The problem of automatic recognition of images or visualized patterns is one that can be effectively solved using neural networks (Bishop, 1995, 2006; Davies, 2003). In Sect. 8.2.1, the applicability of neural network technology to constructing quantitative diagnostic models in complex cases of structural anomaly recognition was mentioned. One aspect of this complexity may be the difficulty of applying classical mathematical methods to identify signs of anomalies. Below, we present a preliminary analysis of the application of neural network technology to the recognition of anomalies whose signs, although obvious, are difficult to express mathematically, in combination with the active vibration diagnostics method.

As shown in Sect. 6.3.4, the calculated vibration displacement amplitudes, which contain information about hidden defects, agree well with the interference pattern obtained experimentally during the testing of this method and properly processed. Thus, both distributions of vibration displacement amplitudes, calculated and experimental, can be considered equivalent from the perspective of pattern recognition. Therefore, when analyzing the applicability of a neural network to the problem in question, we can limit ourselves to the calculated data. These data were obtained for a structural part – a steel branch pipe, which is typically used to connect a pipeline to a pressure vessel. One end of the branch pipe, which has a thicker wall, is rigidly fixed, while the other is considered free. Between the two ends is a weld seam, into which an artificial defect similar to a macrocrack is introduced, as shown in Figure 8.1.

The purpose of the preliminary analysis, conducted by the author together with engineer Igor V. Uspensky (who carried out the calculations), was to find answers to the following questions:

1. How does such a defect affect the frequency spectrum of vibrations?
2. Is it possible to distinguish the distortions in the vibration mode shapes caused by the defect at the background of their changes caused by the geometry of the weld seam?
3. What diagnostic signs can be found to create a quantitative diagnostic model in the form of a neural network?

To conduct the analysis, two finite element models of the branch pipe were constructed: one without the defect and the other with it. The automatically generated finite element mesh included several layers of solid elements across the wall thickness and was sufficiently dense in the area of the defect. Natural frequencies and vibration mode shapes were calculated for each model and then compared with each other.

In answer to the first question, the calculations confirmed the obvious assumption: the presence of a defect that violates the axial symmetry of the structural part under consideration leads to the appearance of non-axisymmetric modes in its vibration spectrum, absent in the defect-free structure. The appearance of such modes can be considered a diagnostic sign, and this is a partial answer to the third question. Comparable vibration modes (which appear both with and without the defect) show minor changes in frequency values – less than 1%.

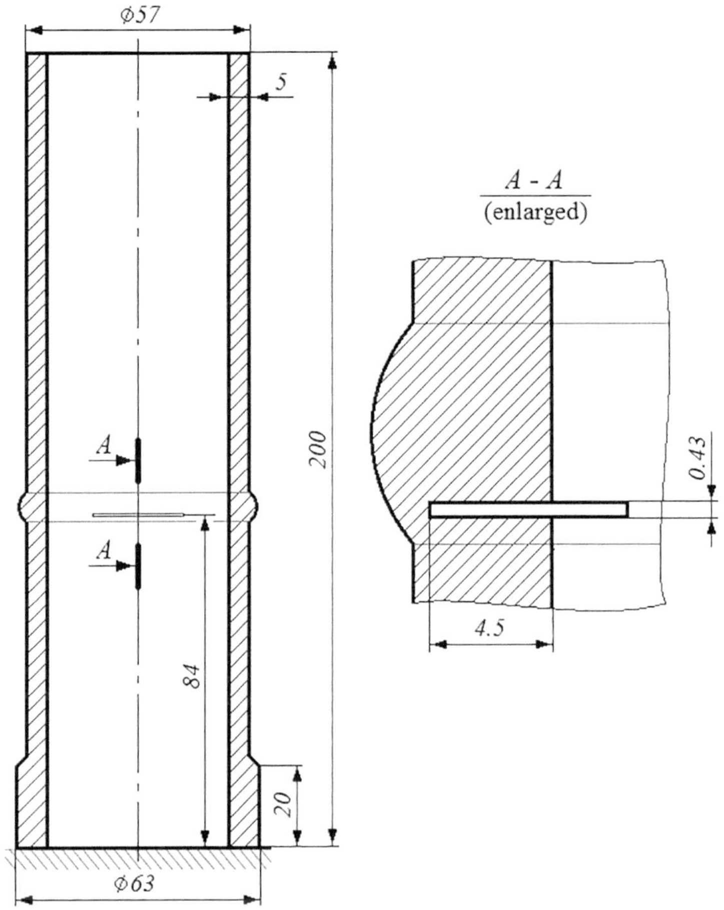

Fig. 8.1 Branch pipe considered in the preliminary analysis of the applicability of a neural network for recognizing hidden defects (linear sizes in mm; circumferential defect size: 60°).

The defect manifests itself in specific distortions of the vibration displacement amplitudes for some of these modes – see Figures 8.2 and 8.3, where the lateral (radial) displacements associated with bending of the branch pipe are presented. We note, as in Sect. 6.3.4, that the distortions are observed on the outer surface of the branch pipe, whereas the defect is located internally, in the weld seam, and is not visible from the outside. This manifestation of a hidden defect is explained by the three-dimensional strain state of the branch pipe wall – due to this, the "opening" of the gap in the wall corresponding to the defect, during bending vibrations, is accompanied by changes in the lateral (radial) displacements of points on the outer surface.

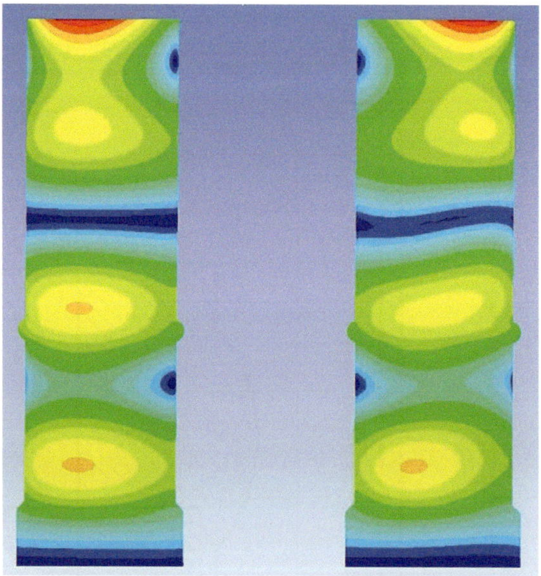

Fig. 8.2 Vibration mode shapes for the branch pipe, 27th vibration mode (bending as a beam): left – without defect (f = 16247 Hz); right – with defect (f = 16241 Hz).

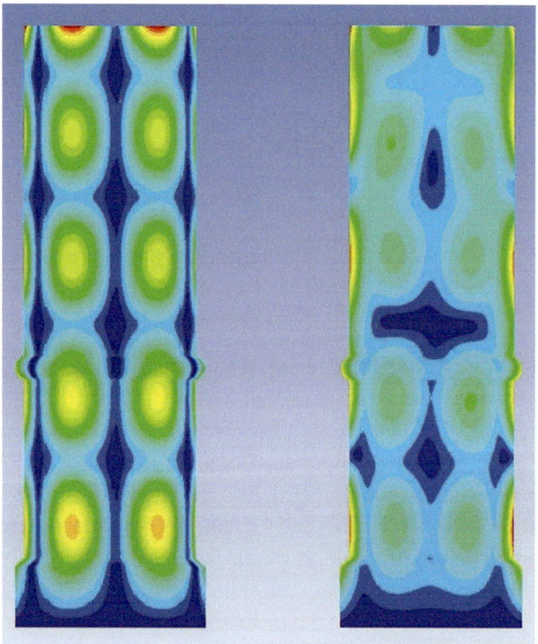

Fig. 8.3 Vibration mode shapes for the branch pipe, 32nd vibration mode (bending as a shell): left – without defect (f = 20096 Hz); right – with defect (f = 20085 Hz).

The distortions represent changes in the contours corresponding to certain levels of vibration amplitudes, including the disappearance of old contours characteristic of a structural part without a defect, and the appearance of new ones due to the presence of a defect. For some vibration modes, these changes are difficult to describe by means of logical structures or mathematical concepts that could be processed within the framework of symbolic AI (Sect. 8.1.1). Therefore, the use of a neural network, which does not require a mathematical formulation of the objects being processed, is preferable.

The obtained calculation results allow for affirmatively answering the second question of our analysis: the distortions in vibration patterns caused by the defect, visible in the vibration displacement amplitude distributions for many modes, are clearly distinguishable at the background of their changes caused by the geometry of the weld seam. At the same time, we obtain a complete answer to the third question: the observed distortions can serve as the main diagnostic signs of hidden crack-like defects detected using the active vibration diagnostics method. The localization of the distortions indicates the location of the defect, and the distortion pattern correlates with the shape and size of the defect. This correlation must be quantified by the neural network itself during its training in accordance with the generally accepted image analysis procedure. When using neural network technology, no formalization of diagnostic signs is required – they should be automatically determined during training. It is advisable to conduct training using examples of images of the calculated vibration mode shapes of branch pipes (or other structural parts) with defects of varying sizes and locations. In this procedure, images should be split into subdomains to form the required number of combinations of input data, $X_1, X_2, \ldots X_d$, and output, Y.

Clearly, this application of AI in the form of a neural network facilitates more efficient detection of hidden structural defects and thus contributes to reducing the uncertainty generated by source 2 in Table 3.1. However, it can be expected that the implementation of defect recognition technique using a neural network is capable of introduce specific uncertainties due to the opacity of the neural network technology itself (Sect. 8.1.2), which is its downside. Defect detection occurs during the post-design stages of the system life cycle, but preparation for the use of the neural network, including its training and the study of the possibility of minimizing this downside, should be part of R&D carried out at the design.

8.3 Control in Critical Situations Created by Humans During Aircraft Flight

8.3.1 Characterization of Anthropogenic Critical Situations in Flight

In Sect. 4.4.1, when considering some applications of analytical methods for reducing uncertainty in the design of technical systems, we noted the insufficient development of a methodology that could take into account the epistemic uncertainty associated with negative manifestations of human factors. These manifestations can lead to critical situations during system operation and require special safety measures. In relation to aircraft design, the potential of AI for solving control problems in critical situations created by humans in flight is of interest. Below, this potential is analyzed by considering flight situations that require replacing human control with automated control based on automatic recognition of the current situation.

In modern aircraft, a significant portion of the flight control functions is performed or can be performed automatically, without the participation of a human operator (pilot). The prospects for civil aviation to transition to fully unmanned transport air vehicles or passenger planes without crews is not clear – it has both advantages and disadvantages (McLean, 2003; Harris, 2003). Without adding further to the discussion of this issue, it should be noted that R&D aimed at reducing human participation in aircraft control is largely motivated by the need to minimize negative manifestations of human factors during flight and the risks associated with them. The range of these negative manifestations covers erroneous actions of the pilot, delayed reaction to rapid changes in weather or air traffic conditions, inattention, fatigue, illness, inaction, suicidal intentions, hijacking of an aircraft by intruders, including terrorists, who are among the passengers, panic behavior of passengers on board, etc. If the escalation of a critical situation caused by any of these initial events – we will call such a situation *anthropogenic* – is not stopped, it can lead to a serious flight accident and even a catastrophe. Some examples of catastrophes of this genesis, which resulted in the destruction of aircraft and human fatalities, are given in Table 8.2.

Table 8.2 Some aviation catastrophes caused by negative manifestations of human factors in flight.

Date	*Location*	*Aircraft, airline and flight*	*Probable[1] cause*
2001-09-11	USA, Washington, DC	Boeing 757-223, *American Airlines*, Flight 77	Hijacking of the aircraft by terrorists (ASN, 2006a)
2001-09-11	USA, PA	Boeing 757-222, *United Airlines*, Flight 93	Hijacking of the aircraft by terrorists (ASN, 2006b)
2001-09-11	USA, New York	Boeing 767-222, *United Airlines*, Flight 175	Hijacking of the aircraft by terrorists (ASN, 2006c)
2001-09-11	USA, New York	Boeing 767-223ER, *American Airlines*, Flight 11	Hijacking of the aircraft by terrorists (ASN, 2006d)
2009-06-01	Atlantic Ocean	Airbus A330-203, *Air France*, Flight 447	Misunderstanding of critical situation by pilots and erroneous actions to get out of it (ASN, 2025c)

Contd...

[1] In the terminology of investigation reports.

Table 8.2 *Contd...*

Date	*Location*	*Aircraft, airline and flight*	*Probable cause*
2013-08-14	USA, BHM	Airbus A300F4-622R, UPS, Flight 1354	Pilot fatigue (ASN, 2013)
2013-11-29	Namibia, Bwabwata National Park	Embraer ERJ-190AR, *Linhas Aéreas de Moçambique* (LAM), Flight 470	The pilot's suicidal intentions (ASN, 2024c)
2015-03-24	France, Prads-Haute-Bléone	Airbus A320-211, *Germanwings*, Flight 9525	The pilot's suicidal intentions (ASN, 2025b)
2016-03-19	Russia, Rostov Airport (ROV)	Boeing 737-8KN, *flydubai*, Flight FZ981	Pilot fatigue (ASN, 2025a; Yakimovich, 2024)
2021-07-06	Russia, Kamchatka, Palana	Antonov An-26B-100, Kamchatka Aviation Enterprise	Lack of time to perform the required maneuver (ASN, 2024b; Yakimovich, 2024)

Fortunately, not every critical situation that arises in flight leads to a catastrophe – sometimes a safe outcome is possible. It is the bifurcation nature that is the distinguishing feature of a critical situation – a control action, applied properly in spatial and temporal aspects, can prevent it from turning into a disaster. This is the difference between a critical situation and an emergency one – the latter, as a rule, is irreversible and results in an accident that is associated with destruction and material losses (one of the emergency situations is an aircraft emergency landing, discussed in Chap. 7). An example of a critical situation created by humans and resolved safely is that which occurred on 17 February 2024, involving *Lufthansa* Airbus A321-231, flight LH1140. It was en route from Frankfurt to Seville. Near Madrid, the captain left the cockpit and shortly after, the co-pilot suddenly lost consciousness. The plane flew without pilot control for about 10 minutes. Upon returning to the cockpit and seeing what had happened, the captain decided to divert to Madrid-Barajas Airport, where he landed without further incident (ASN, 2024a).

Anthropogenic critical situations are characterized by uncertainty of occurrence, development and outcome. They are caused by human factors, indicated as the fifth source of uncertainty in Table 3.1. This uncertainty includes both aleatory and epistemic components. The former primarily encompasses pilot behavior features that are deviations from normal piloting but are predictable and associated with erroneous actions, delayed reactions, inattention, or fatigue. The uncertainty of the occurrence of a critical situation arising from any of these causes can be characterized by an appropriate probability estimate based on statistics accumulated in aviation. Probability estimates of deviations in pilot actions, which are aleatory in nature, can be refined in a similar way to how this is done for stochastic variations in structural properties and maximum loads (Sect. 5.1). Based on the known causes of critical situations of this type and reasonable probability estimates, standard design and organizational measures can be implemented in order to minimize the risk of an accident or catastrophe.

Other negative manifestations of human factors during aircraft flight – sudden illness of the pilot, inaction, suicidal intentions, hijacking, etc. – are rather covered by the category of epistemic uncertainty. They are difficult to predict and almost impossible to control, however, some preventative design and organizational measures are feasible.

Existing approaches to managing predictable anthropogenic critical situations of the first and second types are discussed in the next section.

8.3.2 Existing Approaches to Managing Anthropogenic Critical Situations

Design measures aimed at minimizing the risk associated with aleatory uncertainty in pilot behavior (behavioral features of the first type) include constructing the cockpit interior and the human-machine interface according to ergonomic principles. These principles are established in existing standards and are being developed through ongoing research – some aspects of ergonomic design are discussed in the author's previous book (Spirochkin, 2023). Organizational measures are implemented in accordance with national and international regulatory documents – see, for example, No. 60-FZ (1997); CFR Part 91; ICAO (2020). These measures cover pilot education and training, candidate selection based on professional suitability criteria upon recruitment and subsequent certification, periodic physical and mental health screening (Yakimovich, 2024), pre-flight health checks of pilots, as well as the implementation of safety culture in airlines. Important elements of the latter include the personal responsibility of private and commercial pilots to perform health assessments in accordance with the IMSAFE checklist (FAA, 2023; Kingsky Flight Academy, 2023) and mutual monitoring achieved through the presence of two pilots in the cockpit.

Regarding the second type of negative human behavior features (the potential manifestation of which is characterized by epistemic uncertainty, but is, in principle, predictable), design measures to reduce risk are limited to protection of the cockpit doors from unauthorized entry of intruders or penetration of small arms bullets and fragments of explosive devices with specified parameters (CFR Part 25, § 25.795). Organizational measures include profiling carried out by airport security services to identify intruders, as well as the use of technical means for screening passengers and baggage.

In the event of an anthropogenic critical situation of any origin during aircraft flight, the traditional approach to flight safety, cultivated in civil aviation, is applied. This approach assumes that any in-flight hazards, including those caused by crew error, must be managed through high qualification of pilots and their appropriate training (ICAO, 2020). However, this approach, as well as the design and organizational measures described above, cannot ensure safety across the entire range of possible anthropogenic critical situations. This disappointing conclusion is confirmed by the list of aviation catastrophes in Table 8.2. Moreover, the existing system of ensuring flight safety in civil aviation, although generally reliable, apparently cannot guarantee the prevention of any anthropogenic critical situation or a safe exit from it in the future. Such a guarantee is hampered by the very nature of human factors, which are insufficiently studied and limited in control.

8.3.3 Alternatives

Awareness of the existing limitations in managing possible negative manifestations of human factors in civil aviation motivates the search for an alternative to traditional approaches. The reversibility of critical situations provides a fundamental opportunity to cope with them and ensure a safe outcome.

In situations created by humans in flight, an alternative approach can be implemented through intervention in the process of human control, which has become inadequate, by the automatic part of the human-machine system. The form of intervention can vary depending on the specifics of the situation. For example, if a critical situation is caused by unintentional erroneous actions by the pilot (in particular due to insufficient understanding of the complex flight situation), but these actions do not create an immediate threat of an accident, then the intervention can be "soft": 1) a recommendation to the pilot to correct the error or 2) a slight automatic modification of the manual control to improve the flight (Rogalski, 2010).

In situations of a more serious nature and rapidly developing in a negative direction, their development can be stopped and a safe outcome ensured by blocking dangerous human behavior and transferring aircraft control to automatic mode. This "hard" intervention can also be used in a critical situation of the first category if the pilot did not follow the recommendation and (or) missed the time to correct his errors. The second type of intervention is of primary interest, since it is the most complex and applicable to the most extreme cases. For cases such as an obvious hijacking attempt or an aircraft deviation from its flight plan noticed by the air traffic control center, not attributable to the actions of the authorized flight crew, a number of technical solutions have been proposed – see, for example, Pizzo (1974); Shear (2003); Gleine (2005); Conner (2009). These solutions are typically based on the crew's or air traffic controllers' assessment of the critical situation, as well as their involvement in control, including remote control of the aircraft from the ground. To the author's knowledge, none of the proposed solutions have been implemented in practice.

The idea of situationally replacing human control, which turns out to be insufficiently reliable or even dangerous, with automatic control goes beyond the traditional design and organizational measures used in civil aviation. It only partially overlaps with the alternative technical solutions presented in the aforementioned publications.

An alternative approach based on this idea and presented below is aimed at ensuring safety in the entire range of the anthropogenic critical situations and is more in line with the philosophy of space technology. Given the extreme nature of the physical processes when launching a spacecraft into orbit or re-entering the atmosphere, as well as the complexity of maneuvers in orbit, which limits the control capabilities of the human operator, this philosophy provides for predominantly automatic control. The participation of the human operator in control is only an option, implemented in a limited number of space operations.

For example, not a single cosmonaut (pilot) was on board the Soviet aerospace shuttle *Buran* during its test flight and return to Earth on November 15, 1988. Its landing on the airfield runway was carried out automatically (Brovkin and Kravets, 2014). A similar concept is currently being used to varying degrees in unmanned aerial vehicles, including military ones. However, the approach proposed here differs from fully automatic control of an aircraft. Its essential distinguishing features are as follows:

- the automatic control system has a special component which works simultaneously with the pilot – it monitors not only the flight conditions, but also carries out a kind of supervision over his actions, as well as the situation in the cockpit and passenger cabin from a safety point of view;
- this component intervenes in the control performed by the pilot when a critical situation arises and carries out autonomous actions – up to the complete exclusion of the pilot from the control loop and the implementation of other protective measures.

To ensure this situational takeover of control, an aircraft – plane, helicopter or aerospace vehicle – should be designed as a human-machine system with a high level of robotization. The special component of the automatic control system must be capable to recognize critical situations (in the context under consideration, anthropogenic), determine an automatic control algorithm adequate to a specific situation and aimed at a safe outcome from it, and also perform appropriate control actions. All these functions must be carried out in real flight time. The current level of science and technology makes it possible to implement the component in question as part of an on-board automatic control system.

The initial phase of R&D carried out in this direction with the participation of the author includes an engineering formulation of the problem and an analysis of the possibilities of its solution.

8.3.4 The Problem Statement

The problem of situational takeover of control when an anthropogenic critical situation arises in flight in order to ensure a safe exit from it can be formulated as follows: it is necessary to supplement the on-board automatic control system with a special component that must solve three lower-level problems in real time:

1) situation recognition;
2) determining an algorithm for managing the situation; and
3) implementing this algorithm.

The first problem is similar to that solved during the investigation of an aviation accident post-factum, but differs from it because it must be solved directly in flight, and the result of the solution must be unambiguous – in the form of an algorithm of automatic actions that would ensure a safe exit from the situation. Reliable situation recognition in both cases requires information obtained through several channels: sound recordings, video data, and sensor data characterizing the flight parameters, as well as control actions and the state of the aircraft systems. All sounds in the aircraft cockpit are currently gathered by the cockpit voice recorder (CVR) and sensor data by the flight data recorder (FDR). These recordings assist accident investigators gain a detailed understanding of how a critical situation developed over time and determine its root causes. Video recordings of the cockpit or passenger cabin are not generally made.

Automatic recognition of a situation in real time is possible using electronic processor devices that are capable of identifying certain markers (or patterns) in the time series of data received through each channel, and on this basis forming integrated judgments about the occurrence of a critical situation of a certain nature and its possible development. The solution to this problem (the first of the low-level problems in our list) is, according to available information, in the early stages of R&D in global civil aviation. No serial-produced electronic devices of this kind were found.

The second low-level problem that requires a solution after recognizing an anthropogenic critical situation is to determine the algorithm for managing it in order to achieve a safe outcome (a safe exit from the situation). Such algorithm must provide for the following groups of operations:

- blocking dangerous or potentially threatening actions of pilots and (or) passengers;
- transferring aircraft control to automatic mode; and
- performing the automatic control with a certain spatial scheme of force application and time profile.

Solving this problem may not be easy, given the variety of critical situations that can arise in flight due to the negative traits of human nature. For the second problem, as for the first, no ready-made solutions applicable to the entire range of possible situations were found in the available sources of information. It is reasonable to assume that the algorithm appropriate to a particular situation can be determined by selecting from a pre-developed set of algorithms, taking into account established criteria of applicability and effectiveness.

The selected algorithm must be implemented (the third of the low-level problems in the list above) by an appropriate control loop, including actuators and secondary flight controls capable of generating the necessary control forces. Air vehicles are currently equipped with control loops that use mechanical, hydraulic or electromechanical actuators and deflecting aerodynamic surfaces (rudders, ailerons, elevons, flaps, etc.). These elements can provide the control actions needed to manage some critical situations. However, in the event of failure of the aerodynamic secondary controls, their actuators, or the entire control loop, there are no proven technical solutions that guarantee a safe outcome. In such cases, other design solutions are required.

The lack of ready-made technical solutions for a number of issues indicates either the novelty of this formulation of the problem at hand or the insufficient attention of designers to it. There is an obvious need to develop appropriate practical proposals for solving these issues in order to create the required component of the automatic control system for an aircraft. The proposals should cover solutions to the following tasks:

- implementation of video recording of the situation in the cockpit and passenger cabin;
- development of methods for recognizing anthropogenic critical situations in real time using all three data acquisition channels: CVR, FDR and video;
- an experimental study of the sufficiency of information received through these channels, and in case of insufficiency, a proposal to obtain additional information;
- embodiment of recognition methods in on-board electronic devices (including the choice of the most effective combination of software and hardware);
- development of control algorithms for a conceivable set of anthropogenic critical situations, including all three groups of operations that must be performed in real time, and a technique for selecting the best algorithm for a specific situation using certain selection criteria;
- development of appropriate control loops for the implementation of these algorithms, including actuators and elements generating control forces;
- development of the architecture of an on-board automatic control system for anthropogenic critical situations, and its integration with the flight control systems currently in use or under design.

8.3.5 Analysis of the Solvability of the Problem

The analysis of the solvability of the problem of creating a component of aircraft automatic control system for managing anthropogenic critical situations is carried out by considering the individual tasks within this problem, which are listed at the end of the previous section.

The first of these tasks (implementation of video recording) seems quite simple from a technical standpoint. However, as attempts to equip cockpits with video cameras have shown, pilots are opposed to such measures, and overcoming their resistance can be a significant organizational challenge. The solution lies in the intersection of industrial and organizational psychology and maintaining the safety culture in airlines.

Methods for recognizing anthropogenic critical situations based on available data from CVR and FDR (see the second task in our list) have so far been developed mainly for use in aviation accident investigations. In particular, the mental state of pilots can be determined by analyzing recordings of their conversations in the cockpit (Yakimovich, 2025b). Attempts are also being made to identify abnormal pilot behavior based on data from FDR, which characterizes the deviations of the aerodynamic secondary controls during manual control in a critical situation (Klyuev, 2024). The key issue for the use of these methods in flight conditions is their applicability on board an aircraft and in real time.

Information obtained from the CVR, FDR, and video (if video recording is feasible) may not be sufficient to recognize critical situations caused by certain unhealthy state of the pilot, his inaction, or suicidal intention, especially when such states do not manifest themselves in the form of audible or visible signs. If the pilot is alone in the cockpit, and if he loses consciousness, remains silent, or shows no visible intention, it is extremely difficult to determine what is happening to him based on this information. Given such situations, it may be necessary to conduct experimental studies to find additional sources of information that can fill this gap (the third task). The best source of information is the one that provides a clear picture of the processes in the pilot's mind that initiated the anthropogenic critical situation.

Methods for identifying processes occurring in pilots' minds, which are used by aviation psychologists in accident investigations and professional suitability assessments (Yakimovich, 2024, 2025a, 2025b), are unlikely to be applicable to recognizing critical situations directly in flight. However, these processes inevitably manifest themselves in certain physical, but non-mechanical, phenomena, which, if identified, can be considered diagnostic signs. Such signs may include, for example, specific patterns in the electromagnetic field emitted by the brain. Despite the weak intensity of the electromagnetic field generated by neurophysiological processes, modern technology enable its recording using instrumental methods. It is possible that additional sources of information necessary for reliable recognition of some anthropogenic critical situations in real time should be based on the development of these methods and their implementation in automatic aircraft control systems.

The solution to the fourth problem (embodiment of recognition methods in on-board electronic devices) should include the choice of a combination of software and hardware based on criteria for their seamless integration into the aircraft automatic control system and minimization of the time spent on recognizing a critical situation.

When solving task 5 (development of control algorithms for a conceivable set of anthropogenic critical situations and a technique for selecting the best algorithm for a specific situation), it is necessary to take into account:

- the nature of each critical situation in question and the human behavioral parameters characterizing it;
- variables describing current flight conditions;

- the actual aircraft state in terms of operability and safety (integrity, controllability, degradation of the properties of elements, etc.);
- control forces necessary to implement a safe exit from of the situation;
- available characteristics of control loop elements;
- the changeability of the situation over time and the time reserve available for control actions;
- the emergence of additional loads at a certain combination of spatial scheme of control forces and their time profiles with the dynamic properties of the aircraft;
- limitations imposed by the environment (flight altitude, terrain, meteorological conditions, etc.).

In many respects, this task is similar to the one discussed in Chap. 7, which concerned ensuring safety in an aircraft emergency landing. Therefore, a solution should likely be sought within the SPARS concept, the application of which is demonstrated there. The functions of the component of the aircraft automatic control system in question, intended to provide control in anthropogenic critical situations, fall within the range of functions of the Critical and Emergency Control System (CECS), the description of which is given in in Sect. 7.3.4. At this stage of the analysis, it is difficult to determine whether it should be an autonomous subsystem within CECS or integrated into this system as a set of additional options.

In the first case, the following is required:

- close information interaction between the autonomous subsystem and CECS;
- joint use of a number of on-board equipment elements;
- independent communication between the subsystem and Intelligent Flight Control System (IFCS) mentioned in Sect. 7.2.3.

In the case of full integration, all interactions with on-board equipment elements and other systems within IFCS are carried out at the CECS level.

The subsystem intended to provide control in anthropogenic critical situations will use, among many on-board equipment elements, actuators and elements generating control forces – both aerodynamic secondary flight controls and new ones not yet used in aviation. The latter could include, for example, small solid-fuel rocket engines similar to those applicable in emergency rescue systems (launch escape systems) or soft landing systems for spacecraft. Technical proposals for these elements can be prepared as a result of solving task 6. These proposals, as well as results of solving task 7, which should determine architecture of the subsystem in question and its integration with IFCS, can form the content of the corresponding conceptual design.

8.3.6 Preference for Neural Network Technology

The key features of the problem of creating a subsystem for control in anthropogenic critical situations from the information technology perspective are as follows:

1) the heterogeneity and multiplicity of data required to recognize such a situation and determine an effective control algorithm, i.e., the specific nature of the data to be processed;
2) the continuous updating of these data as the critical situation develops, necessitating its processing and obtaining results in real time;
3) the requirement for installation onboard an aircraft.

We encountered similar features when considering the problem of ensuring safety in emergency landing (see, in particular, Sects. 7.4.3 and 7.5.1). In that case, neural network technology proved to be the most effective in terms of the minimum required computer resources and the time it takes to produce inferences. It also appears preferable for recognizing anthropogenic critical situations onboard an aircraft. Pattern recognition based on characteristic signs is a typical task that can be solved by a neural network (Sect. 8.2.2). Recognition in this case does not require the construction of a complex logical model, which would be inevitable when using symbolic AI. The identification of characteristic signs should be performed by relevant neural network itself during its training. The difficulty lies in selecting a representative set of examples for its reliable training, the results of which will subsequently provide the ability to unambiguously assess a new critical situation. Given the diversity of such situations, training is a challenging task, but it is carried out not in flight, but on the ground and in advance.

Determining control algorithms for various flight situations is possible using classical methods of applied mathematics and mechanics, including solving the differential equations of aircraft motion. However, this requires extensive computational resources unavailable onboard an aircraft and is not applicable under the time constraints typical of a developing critical situation. Using a neural network, pre-trained on a representative set of examples corresponding to a conceivable set of anthropogenic critical situations and flight conditions, ensures the algorithm can be developed with minimal time expenditure.

Another preferred area of application of neural network technology is the processing of the results of experimental study aimed at identifying additional channels of information for the purpose of more reliable recognition of an anthropogenic critical situation. An important argument in favor of neural networks is their ability to process fuzzy data, which is likely to be present in information flows of various types.

It should be added that an aircraft that implements control capabilities in anthropogenic critical situations acquires the properties of a robotic system and can be called "smart" (Spirochkin, 2025). The described application of AI enables designers to cope with uncertainty created by human factors (source 5 in Table 3.1), including its epistemic component. However, along with the minimization of this uncertainty, another one can be expected to emerge, due to the opacity of neural network technology, noted in Sect. 8.1.2. Such "expected" uncertainty may be related, depending on the application, to source 2 (hidden imperfections of the system control loop) or source 6 (inaccuracies in the computational substantiations of the design decisions). Reducing this uncertainty should become one of the R&D areas during the design stage of a "smart" aircraft.

8.4 Potential of Neural Network Technology for Solving Other Design Problems

Thus, we have considered, using practical examples, three potential areas of application of neural networks in designed technical systems: forecasting (Sect. 7.4.3), pattern recognition (Sects. 8.2.2 and 8.3.6), and autonomous control (Sects. 7.4.3, 7.5.1, and 8.3.6). The universal nature of neural network technology, low requirements for the mathematical properties of the information objects to be processed, and the comparative simplicity of implementation determine the potential of this technology for solving other, albeit smaller-scale, design problems. As such problems, we can mention the following:

- detection of characteristic, stable patterns in large data sets, including quantitative relationships between variables;
- substantiation of changes being made to the designed system without additional calculations;
- identification of mathematical models of physical objects based on experimental tests.

These problems are primarily related to the experimental and computational substantiations of design decisions. Such substantiations can generate uncertainty, represented by source 6 in our classification (Table 3.1). Therefore, solving the mentioned problems helps minimize this uncertainty.

Methods for solving the first problem belong to the fields of knowledge discovery, data mining, and data classification (Keogh et al., 2002; Sree and Vardhani, 2015; Infomineo, 2024; Pandey et al., 2025). The findings from processing large arrays of reference information, as well as experimental data, can also help to reveal unknown factors and, thus, minimize the uncertainty arising from source 7, listed in the same table.

Making changes to a technical system under design typically requires repeating calculations to analyze their impact on the safety and performance of the system. Such calculations are particularly necessary when changes affect the properties of load-bearing structural elements or key operating parameters. These calculations require additional labor from analysis engineers and consume the corresponding computing resources. The need for repeated calculations is primarily due to uncertainty in the form of insufficient knowledge of the dependence of safety indicators, particularly the strength and reliability of the structure, on numerous variables that are directly changed or affected by the changes. If these dependencies were precisely known, substantiating the changes would not require complex calculations and would not entail additional costs.

In Sect. 7.4.3, we demonstrated the feasibility of forecasting the dynamic response of an aircraft structure under varying loading conditions and mass for the design case "emergency landing" using a neural network. The basis for this forecasting was the determination of a generalized function $\mathcal{F}$, which links the structural response parameters characterizing safety with a number of input parameters influencing them. It is clear that neural network technology, which makes it possible to make such forecasts, is also applicable to solving the second problem listed above, namely, substantiating design changes without performing additional calculations each time.

Identification of mathematical models used in the design and operation of a technical system is possible based on the results of experimental tests of physical objects, which can provide information on the actual properties of the system. Sources of such information can be tests conducted at intermediate design stages using small-scale physical models, system and certification tests, or special calibration (model updating) experiments (Oberkampf et al., 2003) conducted at the end of the design process using experimental or standard system specimens. Updated and refined (identified) mathematical models are applicable to automated control during operation, in-service monitoring and diagnostics, as well as other life cycle operations.

If we limit ourselves to the area of parametric identification of structural models (i.e., refining the parameters of a mathematical model whose formal construction is known), then its methodology is similar to that used in the diagnostics of structural anomalies (Sect. 8.2.1). The experimental data on which such identification is based are often the results of modal tests or other similar approaches (Van der Auweraer and Hermans, 1999; Kon and Zhao, 2007; Evins, 2009; Feng et al., 2020). The applicability and preference of neural network technology in processing modal data were discussed in Sect. 8.2.2.

To identify models constructed using FEM, two alternative techniques are applicable: refining the members of the stiffness, inertia, and damping matrices of the finite element ensemble, or updating the elastic, inertial, and dissipative properties of individual finite elements. The second technique appears preferable from a physical point of view. It allows the analysis engineer to monitor (track) and evaluate identification results using physical feasibility criteria. This is particularly important given the opacity (potential incomprehensibility) of neural network output.

8.5 On the Risks Associated with AI

8.5.1 Potential Harm from Transferring Functions Inherent in Human Intelligence to AI

Within the flow of publications on the topic of AI, there is a "current" that presents criticism of the uncontrolled introduction of AI into human life and activity, as well as a discussion of the threats this process poses (Saraçyakupoğlu, 2020; Bailey, 2023; McKinsey Global Institute, 2023; NIST, 2023; Copeland, 2025). We set aside concerns regarding potential privacy violations associated with the hidden collection of personal data when users access various information resources and publicly available AI tools. Let us focus on those components of potential harm that have a broader impact on society and (or) are relevant to the safety of technical systems. According to existing critical judgments, the potential harm from passing functions inherent in human intelligence to "intelligent" computer systems consists of three main components:

1) many people could lose their jobs;
2) human cognitive abilities will degrade without proper application;
3) the processes for obtaining results using AI are not transparent and therefore difficult to verify, which is especially dangerous in areas related to safety.

The history of the implementation of any machine technology, including information technology, shows that despite the reduction in employment when humans are replaced by machines, a catastrophic collapse of the labor market does not occur; on the contrary, new employment opportunities open up. The use of computers has led to the expansion of rare earth element mining and processing, as well as the emergence of entire industries, such as the production of electronic microchips, processors, and other hardware components, programming, digital communications, data security, and so on. Consequently, the first component of potential harm can only be localized and temporary. In developed socioeconomic systems, it can be minimized.

The situation is worse with the degradation of human cognitive abilities. In accordance with some studies, the highest level of intellectual development of mankind occurred in the period from the end of the XIX century to the 60s of the XX century. Since the 1970s, the growth of such an intelligence indicator as IQ (intelligence quotient) has slowed down, and then stopped, if it did not go backwards (Sundet et al., 2004; Flynn, 2009). Part of the community of neurophysiologists and psychologists believes that humanity is beginning to "get stupid" because it increasingly relies on "electronic brains" to perform tasks previously performed by the human brain, which, left without exercise, degrades (Kurpatov, 2018, 2020). In our time, when any information can be obtained through the Google search engine, almost nothing needs to be memorized, learned independently from one's own experience and analyzed, most people prefer ready-made answers. To be fair, other opinions should also be noted. For example, some researchers doubt that the increase in IQ observed in previous decades reflected a genuine increase in intelligence, and was not an artifact or a consequence of improved nutrition, school education and other living conditions (Neisser, 1997). Others suggest that progressive air pollution, which currently affects more than 90% of the world's population, is a possible reason for the decline in intelligence observed worldwide (Schraufnagel et al., 2019).

In any case, regardless of which explanation is correct, what appears to be a degradation in human cognitive abilities coincides in time with the expansion of AI presence in our lives and with the transfer of an increasing number of intellectual functions to it. The power of information technology in general, and existing AI elements in particular, already significantly exceeds human capabilities in terms of data volume and processing speed. Any superior power is alarming, because as if it gets out of control, it threatens unpredictable consequences.

In Sects. 7.4.3 and 8.3.6, we concluded that neural network technology is preferable for use in critical and emergency situation control. However, along with these opportunities come specific risks. These are associated with the indirect danger of powerful automated tools: by relying solely on them, humans do not develop the ability to act independently and adequately when unexpected threatening circumstances arise. Yet this ability is inherent to humans as biological beings. It characterizes the plasticity of their consciousness and is aimed at ensuring survival.

For example, research in the field of aircraft cockpit automation has shown that transferring control to automated systems leads to excessive trust and complacency in pilots, a decrease in their ability to exercise situational awareness, and a lack or loss of previously acquired experience in critical and emergency situations (Grindley et al., 2024). This trend is not related to the degradation of human cognitive abilities, but rather should be associated with the third component of potential hazards – the risks associated with the use of AI in the control of safety-critical systems when human control capabilities are limited.

As noted earlier, the advantages of using AI in automated in-service monitoring and diagnostic systems, as well as in control loops for technical systems are the abilities to:

- detect signs of impending adverse events in large volumes of constantly changing data;
- process fuzzy information and self-learn;
- find solutions for managing critical and emergency situations in real time; and
- implement control that is ahead of their dangerous development.

However, when using neural networks, the processes of obtaining outputs, selecting control algorithms, and subsequent automatic actions aimed at ensuring safety may be incomprehensible to humans. The changeability of a neural network during training makes it difficult to understand how it will make decisions in a given situation and what the consequences will be. Therefore, at the current stage, the use of AI for automatic control in critical situations with potentially catastrophic outcomes raises reasonable doubts.

Thus, AI elements designed to solve tasks in which their capabilities exceed those of humans and are therefore highly desirable create additional uncertainty. This uncertainty is exacerbated by human interaction with a machine equipped with an "intelligent" control system containing AI elements, when the operator has little understanding of how the system works. The danger of such situations was demonstrated by two crashes of Boeing 737-8 (MAX) – near Jakarta in 2018 and near Addis Ababa in 2019, which were discussed in Sect. 1.2.2.

Another aspect to note is that since AI elements are still created by humans, they may inherit human errors made during their creation.

Concerns about the growing role of "intelligent" machines against the backdrop of the supposed decline of human cognitive abilities are intensified when

AI is "entrusted" with the most subtle and creative mental work, the prerogative of scientists: the acquisition of new knowledge. Neural networks, which search for patterns in accumulated data, are increasingly being used for this purpose. Generative AI, based on identified statistical correlations, is becoming the most widely used tool for the creation of new intelligent products (texts, images, software algorithms, etc.). This threatens to devalue the human faculty for intuition, which is precisely what enables discoveries and scientific and technological breakthroughs in unexplored areas.

The possibility of human errors in the development and training of neural networks, on the one hand, and the correlations detected by neural networks without understanding the cause-and-effect relationships between variables and critical analysis (which the human intellect is capable of), on the other hand, create significant risks when equipping safety-critical technical systems with AI elements.

8.5.2 Increasing Resource Consumption, and Environmental Impact

Growing demand for high-performance processors that power AI has spurred a boom in microchip factory construction around the world. Each new generation of microchips inside such processors becomes more complex than the previous one, requiring more energy, materials including rare earth elements, and water for manufacturing. For example, a modern semiconductor manufacturing plant requires approximately one gigawatt of electricity and can consume about 100,000 tons of water per day (Alexander, 2025). When most electricity is generated using coal and gas, the operation of such a plant is associated with high carbon dioxide emissions. Gases used in semiconductor manufacturing, such as sulfur hexafluoride, which can be released and accumulate in the atmosphere, have an even more significant negative impact on the climate than carbon dioxide. In addition, these industrial activities generate highly toxic waste, including large amounts of per- and polyfluoroalkyl substances (PFAS), which do not decompose naturally, causing long-term pollution. In many cases, manufacturers ignore such risks, resulting in adverse consequences for both workers and local communities. The strategy of unlimited expansion of the semiconductor industry is supported by governments, which consider microchip production critical to national security and global competitiveness.

It is also worth noting that the mining of the rare earth elements and other minerals crucial to the production of semiconductors is expanding largely due to its organization in remote regions of the planet, characterized by particularly fragile ecosystems, many of which are also carbon dioxide sinks.

The production of more powerful graphics processing units (GPUs) that run AI and use the latest microchips, is accompanied, like semiconductor manufacturing, by increased resource consumption and a negative environmental impact. For example, carbon dioxide emissions from Nvidia's GPU production increased by 87% in 2024 (ibid).

Training neural networks developed by major AI players can involve adjusting billions of weighting coefficients (Sect. 8.1.2) through repeated calculations, requiring enormous computing power. This process is implemented using thousands of GPUs and specialized tensor processing units (TPUs), as well as central processing units (CPUs), all working in parallel. Training can take considerable time and consume a huge amount of electricity. Computers with arrays of these processors and the telecommunications equipment that provides access to neural networks are housed in data centers, the number of which is also rapidly growing. Some of the major players are setting goals to build AI infrastructure with capacities in the hundreds of gigawatts. According to forecasts, by 2030–2035, data centers could account for approximately 20% of global electricity consumption (Kandemir, 2025), resulting in enormous strain on the power grids and reduced quality of electricity supply in the residential sector. Pressure on water resources is already increasing as computing equipment requires very large volumes of water to cool.

The short service life of GPUs and other high-performance computing components is leading to a growing problem of electronic waste, for which adequate recycling technologies have not yet been developed. Often, aged, obsolete, or failed hardware is simply discarded, polluting the environment.

Thus, the rapid development of AI in its current incarnation is gradually turning into the largest infrastructure project on the planet, accompanied by an unprecedented increase in resource consumption and an increasing burden on the environment. It can significantly alter the energy and environmental balance in the near future. As executives at the world's largest information technology companies ponder the computing power required for the development of AI, questions arise about how such development fits into the planet's existing natural resources and whether it will lead to fatal environmental destruction.

The paradox is that the natural intelligence inherent in humans and perhaps other living beings, based on protein structures and the electrochemical processes within them (which are still poorly understood), requires incomparably less energy and undoubtedly has greater creative capabilities than existing AI elements created from semiconductor circuits.

8.5.3 On Minimizing Risks when Implementing AI

Thus, AI provides designers with significant opportunities to minimize uncertainty generated by a number of sources, including opportunities to reduce epistemic uncertainty. However, other, specific uncertainties arise due to the opacity of AI operations. Therefore, an area of AI development known as *explainable artificial intelligence*, which aims to enable humans to understand how AI produces its outputs, is becoming increasingly relevant (EDPS, 2023).

Even a small degree of inexplicability, as well as the possibility of human error being reflected in AI elements, creates the risks noted in Sect. 8.5.1. Risks are also caused by the avalanche-like development of AI, which is insufficiently controlled, and these risks, as shown in Sect. 8.5.2, can be existential in nature.

Yet, the risks associated with integrating AI into technical systems are essentially typical of any new technology in the early stages of its development, when uncertainty is high and the probabilities of various outcomes are impossible to estimate. To minimize risk in such circumstances, the approach to separating (decoupling) uncertainty and risk, mentioned in Sect. 2.4.1 and referring to the publication by Cocron and Aronhime (2022), appears useful.

Following this approach, we display the relationship between the uncertainty of the outcome of innovation activities and risk, understood as an indicator of potential harm (material or financial losses for stakeholders, adverse consequences for consumers, society as a whole or the environment), on a plane, dividing it into four areas shown in Figure 8.4 and corresponding to intuitively understandable combinations:

1) small uncertainty – low indicator of harm;
2) great uncertainty – low indicator of harm;
3) small uncertainty – high indicator of harm; and
4) great uncertainty – high indicator of harm.

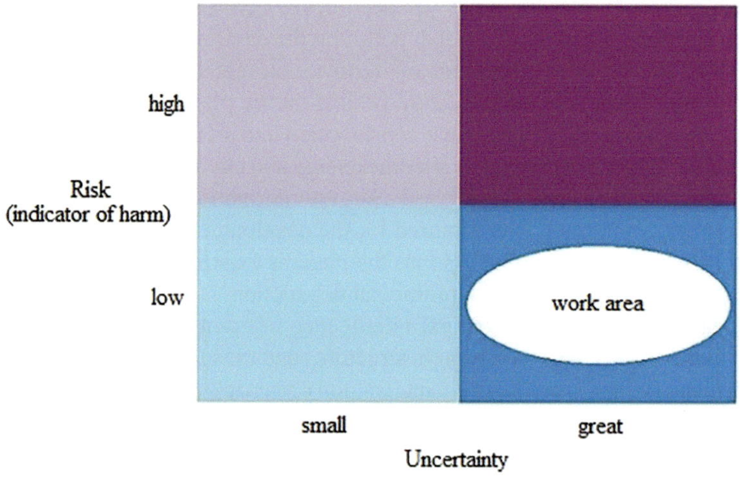

Fig. 8.4 Combinations of uncertainty and risk (indicator of harm) in innovation activities.

When integrating AI into technical systems, especially those critical to safety, the uncertainty of the outcome of innovation is certainly "great". However, AI developers, system element manufacturers, the construction organization and operator (operating organization), and other stakeholders, as well as the regulatory body representing the interests of society as a whole, can choose the indicator of potential harm: it can be reduced by limiting the implementation of "intelligent" devices in the system and "intelligent" functions in its control loop. Under these conditions, it is preferable to carry out innovation-related work in an area where "great" uncertainty (which is actually "given") is combined with "low" harm (or risk, which is the subject of choice). After choosing such a "work area" (see Figure 8.4) and implementing the innovation, it is possible to obtain data on its success or failure during the trial operation of the created system, and, if necessary, corrective measures can be applied.

One variant of the described approach is the phased implementation of AI, which involves testing its initial application in lower-risk industries such as the automotive or household goods industries (Saraçyakupoğlu, 2020). But the possibilities of even such a cautious technical policy are limited, because according to Amara's Law: "We tend to overestimate the effect of a technology in the short run and underestimate the effect in the long run" (Lin, 2024).

It is noteworthy that Amara's Law, which refers to the specific manifestation of uncertainty in the introduction of new technologies, returns us to the general characterization of uncertainty presented at the very beginning of this book, in Sect. 1.1.1. This phenomenon is subjective, meaning uncertainty exists not outside of us, but within us – in our understanding of the world. All the above methods aimed at minimizing uncertainty are effective only to the extent that they increase our awareness of the situation in which we find ourselves. They do not directly change the world around us, which we perceive as full of uncertainty, although they do influence it through the human-created artifacts in which these methods are implemented (unfortunately, this influence is sometimes destructive). In most cases, uncertainty minimization methods provide a more complete and accurate understanding of this world and thereby provide the opportunity to improve safety.

References

Alexander, A. (2025) The ecological cost of AI is much higher than you think. *Truthdig*, November 10, 2025. https://www.truthdig.com/articles/the-ecological-cost-of-ai-is-much-higher-than-you-think. Accessed November 16, 2025.

Annepaka, Y. and Pakray, P. (2025) Large language models: a survey of their development, capabilities, and applications. *Knowledge and Information Systems*, Vol. 67, pp. 2967–3022.

Arnold, V.I. (1958) O predstavlenii funktsii neskolkikh peremennykh v vide superpozitsii funktsii menshego chisla peremennykh (On the representation of functions of several variables as a superposition of functions of a smaller number of variables). Matematicheskoe prosveschenie (*Mathematical education*), Series 2, Issue 3, pp. 41–61. https://www.mathnet.ru/mp490 (in Russian).

ASN (2006a) Aviation Safety Network – ASN Aviation Safety WikiBase. https://asn.flightsafety.org/wikibase/323226. Accessed May 14, 2025.

ASN (2006b) Aviation Safety Network – ASN Aviation Safety WikiBase. https://asn.flightsafety.org/wikibase/323227. Accessed May 14, 2025.

ASN (2006c) Aviation Safety Network – ASN Aviation Safety WikiBase. https://asn.flightsafety.org/wikibase/323228. Accessed May 14, 2025.

ASN (2006d) Aviation Safety Network – ASN Aviation Safety WikiBase. https://asn.flightsafety.org/wikibase/323229. Accessed May 14, 2025.

ASN (2013) Aviation Safety Network – ASN Aviation Safety WikiBase. https://asn.flightsafety.org/wikibase/320696. Accessed May 14, 2025.

ASN (2024a) Aviation Safety Network – ASN Aviation Safety WikiBase. https://asn.flightsafety.org/wikibase/352536. Accessed September 15, 2025.

ASN (2024b) Aviation Safety Network – ASN Aviation Safety WikiBase. https://asn.flightsafety.org/wikibase/318975. Accessed May 17, 2025.

ASN (2024c) Aviation Safety Network – ASN Aviation Safety WikiBase. https://asn.flightsafety.org/wikibase/320634. Accessed May 14, 2025.

ASN (2025a) Aviation Safety Network – ASN Aviation Safety WikiBase. https://asn.flightsafety.org/wikibase/320111. Accessed May 17, 2025.

ASN (2025b) Aviation Safety Network – ASN Aviation Safety WikiBase. https://asn.flightsafety.org/wikibase/320334. Accessed May 14, 2025.

ASN (2025c) Aviation Safety Network – ASN Aviation Safety WikiBase. https://asn.flightsafety.org/wikibase/321502. Accessed May 14, 2025.

Bailey, M. (2023) How can we trust AI if we don't know how it works. *Scientific American*. The Conversation US in October 3, 2023. https://www.scientificamerican.com/article/how-can-we-trust-ai-if-we-dont-know-how-itworks. Accessed November 15, 2025.

Bishop, C.M. (1995) *Neural networks for pattern recognition*. Clarendon Press, Oxford.

Bishop, C.M. (2006) *Pattern recognition and machine learning*. Springer, Singapore.

Brovkin, A.G. and Kravets, V.G. (2014) Automatic landing of the unmanned reusable orbiter *Buran*. *Space Engineering and Technology*, No. 1(4), pp. 75–85 (in Russian).

CFR Part 25. Code of Federal Regulations/Title 14/Chapter I/Subchapter C/Part 25 – Airworthiness Standards: Transport Category Airplanes. https://www.ecfr.gov/current/title-14/chapter-I/subchapter-C/part-25. Accessed July 11, 2025.

CFR Part 91. Code of Federal Regulations/Title 14/Chapter I/Subchapter F/Part 91 – General Operating and Flight Rules. https://www.ecfr.gov/current/title-14/chapter-I/subchapter-F/part-91. Accessed September 20, 2025.

Cocron, A. and Aronhime, L. (2022) Risk, uncertainty and innovation. *NATO Review*. Opinion, analysis and debate on security issues, 14 April 2022. https://www.nato.int/docu/review/articles/2022/04/14/risk-uncertainty-and-innovation/index.html. Accessed May 30, 2025.

Conner, J.P. (2009) Emergency flight control system. Patent US 7,568,662 B1. August 4, 2009.

Copeland, B.J. (2025) Artificial intelligence. Encyclopaedia Britannica, November 4, 2025. https://www.britannica.com/technology/artificial-intelligence. Accessed November 8, 2025.

Davies, E.R. (2003) Visual inspection, automatic (robotics). In R.A. Meyers (ed.) *Encyclopedia of Physical Science and Technology* (Third edition). Academic Press, Cambridge, MA, pp. 489–508.

EDPS (2023) TechDispatch: Explainable artificial intelligence. Technology and Privacy Unit of the European Data Protection Supervisor. European Union. https://www.edps.europa.eu/system/files/2023-11/23-11-16_techdispatch_xai_en.pdf. Accessed December 31, 2025.

Enríquez, J.G., Morales-Trujillo, L., Calle-Alonso, F., Domínguez-Mayo, F.J. and Lucas-Rodríguez, J.M. (2019) Recommendation and classification systems: a systematic mapping study. *Scientific Programming*, Vol. 2019, 8043905, pp. 1–18.

Evins, D.J. (2009) *Modal testing – theory, practice and application*. 2nd edn. Research Studies Press Ltd., Baldock, Hertfordshire, England.

FAA (2023) FAASTeam. Safer skies through education. Are you fit to fly? Federal Aviation Administration. https://www.faa.gov/sites/faa.gov/files/2023-10_Fit_to_Fly.pdf. Accessed September 20, 2025.

Feng, Z., Ye, Z., Wang, W., Lin, Y., Chen, Z. and Hua, X. (2020) Structural model identification using a modified electromagnetism-like mechanism algorithm. *Sensors*, 20(17):4789.

Flynn, J.R. (2009) Requiem for nutrition as the cause of IQ gains: Raven's gains in Britain 1938-2008. *Economics & Human Biology*, 7(1):18–27.

Frischkorn, G.T. and Schubert, A.L. (2018) Cognitive models in intelligence research: advantages and recommendations for their application. *Journal of Intelligence*, 6(3):34.

Garnelo, M. and Shanahan, M. (2019) Reconciling deep learning with symbolic artificial intelligence: representing objects and relations. *Current Opinion in Behavioral Sciences*, Volume 29, pp. 17–23.

Gleine, W. (2005) Aircraft anti-terrorism security system. Patent US 6,844,817 B2, January 18, 2005.

Goodfellow, I., Bengio, Y. and Courville, A. (2016) *Deep learning*. MIT Press, Cambridge, MA.

Grindley, B., Parnell, K.J., Cherett, T., Scanlan, J. and Plant, K.L. (2024) Understanding the human factors challenge of handover between levels of automation for uncrewed air systems: a systematic literature review. *Transportation Planning and Technology*, 48(6), 1383–1408.

Harris, D. (2003) The human factors of fully automatic flight. *Measurement + Control*, 36(6):184–187.

Infomineo (2024) Data mining explained: the art and science of discovering patterns. *Infomineo*, November 21, 2024. https://infomineo.com/services/data-analytics/data-mining-explained-the-art-and-science-of-discovering-patterns. Accessed December 28, 2025.

ICAO (2020) Doc 9868. *Procedures for air navigation services. Training*. Third edition. International Civil Aviation Organization, Montreal, Canada.

Kandemir, M. (2025) Why AI uses so much energy – and what we can do about it. PennState. Institute of Energy and Environment, April 8, 2025. https://iee.psu.edu/news/blog/why-ai-uses-so-much-energy-and-what-we-can-do-about-it. Accessed November 17, 2025.

Keogh, E.J., Lonardi, S. and Chiu, B.Y. (2002) Finding surprising patterns in a time series database in linear time and space. *Proceedings of the Eighth ACM SIGKDD International Conference on Knowledge Discovery and Data Mining*. Association for Computing Machinery, New York, pp. 550–556.

Kingsky Flight Academy (2023) IMSAFE Checklist: how pilots assess their health. Kingsky Flight Academy, January 23, 2023. https://www.kingskyfa.com/post/imsafe-checklist-how-pilots-assess-their-health. Accessed September 15, 2025.

Klyuev, A.V. (2024) Uroki aviakatastrof (Lessons from aviation catastrophes). YouTube. https://www.youtube.com/watch?v=XyOA3nhDYe0. Accessed September 17, 2025 (in Russian).

Kolmogorov, A.N. (1956) O predstavlenii nepreryvnykh funktsii neskolkikh peremennykh superpozitsiyami nepreryvnykh funktsii menshego chisla peremennykh. (On the representation of continuous functions of several variables by superpositions of continuous functions of a smaller number of variables). Doklady Akademii Nauk SSSR (*Proceedings of the USSR Academy of Sciences*), 108(2):179–182 (in Russian).

Kolmogorov, A.N. (1957) O predstavlenii nepreryvnykh funktsii neskolkikh peremennykh v vide superpozitsii nepreryvnykh funktsii odnogo peremennogo i slozheniya (On the representation of continuous functions of several variables by superpositions of continuous functions of one variable and its addition). Doklady Akademii Nauk SSSR *(Proceedings of the USSR Academy of Sciences)*, 114(5):953–956 (in Russian).

Kon, C.G. and Zhao, S.L. (2007) A hybrid strategy for parameter identification of plated structures. In N.E. Shanmugam, C.M. Wang (eds.) *Analysis and design of plated structures*. Woodhead Publishing, Volume 1 in Woodhead Publishing Series in Civil and Structural Engineering, pp. 422–444.

Kurpatov, A.V. (2018) Neironauka o narastayuschei gluposti (Neuroscience about increasing stupidity). *Fontanka.ru*, 15.01.2018. https://www.fontanka.ru/2018/01/15/060. Accessed November 15, 2025 (in Russian).

Kurpatov, A.V. (2020) Vystuplenie v Davose na biznes-zavtrake Sberbanka (Speech at the Sberbank business breakfast in Davos). YouTube, 29 January 2020. https://www.youtube.com/watch?v=GbLclnu-QGc. Accessed November 15, 2025 (in Russian).

Lai, M.-J. and Shen, Z. (2025) The Kolmogorov superposition theorem can break the curse of dimensionality when approximating high dimensional functions. https://arxiv.org/abs/2112.09963. Accessed January 24, 2026.

Lin, P. (2024) Amara's law and its place in the future of tech. IEEE Computer Society, September 6, 2024. https://www.computer.org/publications/tech-news/trends/amaras-law-and-tech-future. Accessed November 17, 2025.

Lorentz, G.G. (1962) Metric entropy, widths, and superpositions of functions. *The American Mathematical Monthly*, 69(6):469–485.

McKinsey Global Institute (2023) Generative AI and the future of work in America. July 2023. McKinsey Center for Government. https://www.mckinsey.com/mgi/our-research/generative-ai-and-the-future-of-work-inamerica. Accessed November 15, 2025.

McLean, D. (2003) Fully automatic flight. An introduction. *Measurement + Control*, 36(6):171.

Minaee, S., Mikolov, T., Nikzad, N., Chenaghlu, M., Socher, R., Amatriain, X. and Gao, J. (2024) Large language models: a survey. arXiv preprint. https://arxiv.org/abs/2402.06196. Accessed February 27, 2025.

Neisser, U. (1997) Rising scores on intelligence tests. *American Scientist*, 85(5):440–447.

Neupane, D., Bouadjenek, R.M., Dazeley, R. and Aryal, S. (2025) Data-driven machinery fault detection: a comprehensive review. *Neurocomputing*, Vol. 627, 129588.

Nielsen, M.A. (2015) *Neural networks and deep learning*. Determination Press.

NIST (2023) Artificial Intelligence risk management framework (AI RMF 1.0). NIST AI 100-1. National Institute of Standards and Technology, U.S. Department of Commerce. https://nvlpubs.nist.gov/nistpubs/ai/NIST.AI.100-1.pdf. Accessed November 15, 2025.

No. 60-FZ (1997) 'Vozdushnyi Kodeks Rossiiskoi Federatsii' ot 19.03.1997 No. 60-FZ (Air Code of the Russian Federation), with amendments and additions, entered into force on November 26, 2025. https://www.consultant.ru/document/cons_doc_LAW_13744. Accessed November 30, 2025 (in Russian).

Oberkampf, W.L., Trucano, T.G., and Hirsch, C. (2003) Verification, validation, and predictive capability in computational engineering and physics. Report SAND2003-3769. Sandia National Laboratories, Albuquerque, NM, and Livermore, CA.

Pandey, S.K., Chand, S., Horkoff, J., Staron, M., Ochodek, M. and Durisic, D. (2025) Design pattern recognition: a study of large language models. *Empirical Software Engineering*, 30:69.

Pizzo, G.A. (1974) Anti hijacking system for aircraft. United States Patent 3,811,643. May 21, 1974.

PNAE G-7-002-86 (1989) Normy raschyota na prochnost oborudovaniya i truboprovodov atomnykh energeticheskikh ustanovok (*Rules for strength analysis of equipment and pipelines of nuclear energy facilities*). Gosatomenergonadzor of the USSR. Energoatomizdat, Moscow (in Russian)

Rogalski, T. (2010) The idea of a system increasing flight safety. *Aviation*, 14(4):112–116.

Roth, E.M., Woods, D.D. and Pople, H.E. Jr. (1992) Cognitive simulation as a tool for cognitive task analysis. *Ergonomics*, 35(10):1163–1198.

Saraçyakupoğlu, T. (2020) The adverse effects of implementation of the novel systems in the aviation industry in pursuit of Maneuvering Characteristics Augmentation System (MCAS). *Journal of Critical Reviews*, 7(11):2530–2538.

Shear, D. (2003) Anti-hijacking system for airplanes and airports. Patent Application Publication. Pub. No.: US 2003/0090382 A1. Pub. Date: May 15, 2003.

Spirochkin, Y. (2023) *Human factors and design*. Springer, Singapore.

Spirochkin, Y. (2025) "Smart" aircraft: control in critical situations created by humans in flight. *Journal of Advance Research in Mechanical and Civil Engineering*, 10(1):13–21.

Schraufnagel, D.E., Balmes, J.R., Cowl, C.T., De Matteis, S., Jung, S.H., Mortimer, K., Perez-Padilla, R., Rice, M.B., Riojas-Rodriguez, H., Sood, A., Thurston, G.D., To, T., Vanker, A. and Wuebbles, D.J. (2019) Air pollution and noncommunicable diseases: A review by the Forum of International Respiratory Societies' Environmental Committee, Part 2: Air pollution and organ systems. *Chest*, 155(2):417–426.

Sree, Y.U. and Vardhani, P.R. (2015) Pattern finding in large datasets with big data analytics mechanism. *International Journal of Computer Engineering in Research Trends*, 2(5):359–364.

Sundet, J,. Barlaug, D. and Torjussen, T. (2004) The end of the Flynn effect?: A study of secular trends in mean intelligence test scores of Norwegian conscripts during half a century. *Intelligence*, 32(4):349–362.

Thomason, R. (2024) Logic-based artificial intelligence. *The Stanford Encyclopedia of Philosophy* (Spring 2024 Edition), E.N. Zalta & U. Nodelman (eds.). https://plato.stanford.edu/archives/spr2024/entries/logic-ai. Accessed November 19, 2025.

Turing, A.M. (1950) Computing machinery and intelligence. *Mind*, 49(236):433–460.

Van der Auweraer, H. and Hermans, L. (1999) Structural model identification from real operating conditions. *Sound and Vibration*, 33(1):34–41.

Van Harmelen, F., Lifschitz, V. and Porter, B. (eds.) (2008) *Handbook of knowledge representation*. Evsevier, Amsterdam, The Netherlands (Series: Foundations of Artificial Intelligence).

Yakimov, V. and Maltsev, G. (2022) Hybrid network structures and their use in diagnosing complex technical systems. *Informatics and Automation*, 21(1):126–160 (in Russian).

Yakimovich, N.V. (2024) *Pilot errors: psychological causes and their prevention*. Izdatelstvo Editus, Moscow (in Russian).

Yakimovich, N.V. (2025a) An objective method for recognizing psychological problems in pilots. *Design Problems and Knowledge Management*, 3(1):7–12. https://saneorg.wixstudio.com/index/dpkm. Accessed September 17, 2025.

Yakimovich, N.V. (2025b) Recognition of mental states in pilots during flights. *Design Problems and Knowledge Management*, 3(2):27–33. https://saneorg.wixstudio.com/index/dpkm. Accessed September 17, 2025.

Yang, X. and Zhu, C. (2024) Industrial expert systems review: a comprehensive analysis of typical applications. *IEEE Access,* Vol. 12, pp. 88558–88584.

Ying, J., Lin, H., Yue, C., Chen, Y., Xiao, C., Shi, Q., Liang, Y., Yau, S.-T., Zhou, Y. and Ma, J. (2025) A neural symbolic model for space physics. *Nature Machine Intelligence,* Vol. 7, pp. 1726–1741.

Zhao, Y. and Smidts, C.S. (2021) CMS-BN: a cognitive modeling and simulation environment for human performance assessment (parts 1 and 2). *Reliability Engineering & System Safety,* Vol. 213, 107775, 107776.

Zwass, V. (2025) Neural network. Encyclopaedia Britannica, October 3, 2025. https://www.britannica.com/technology/neural-network. Accessed November 19, 2025.

9 Changes in Engineering Design over Time, Their Impact on Uncertainty, and New Challenges

This chapter provides a bird's eye view of the engineering design landscape, showing its evolution over time. The changes in the types of technical systems being created, the design methodologies being applied, and the regulatory approaches governing design activities are considered. The analysis focuses on the specific impact of changes occurring in each of these categories on the uncertainty that remains an inevitable phenomenon in design. New challenges are also considered. It is hoped that advanced methods of dealing with uncertainty, including those currently under development, will ensure its minimization, even in the face of emerging problems.

9.1 The Changing Landscape of Engineering Design

Engineering design is a developing area of human activity. Development occurs both through the improvement of existing technical systems, including the replacement of older generations with new ones, and through the creation of entirely new types of machines, industrial and infrastructure facilities. The drivers of development are generally considered to be societal needs for energy, transportation and communication, safety, and understanding of the world around us. Typically, reaching a certain technological level makes it possible to produce more efficient systems with less resource consumption, generate more energy with less expense, travel faster with less fuel consumption in more reliable vehicles, transmit and process larger amounts of data at higher speeds, and create increasingly complex and advanced tools for fundamental research on Earth and in space.

In addition, the following development factors should be noted: competition between technologically leading countries and between companies creating high-tech products, as well as the personal qualities possessed by individuals with engineering talent – the vision, determination, and energy needed to spark interest in others and attract the necessary resources. Such individuals are capable of generating breakthrough ideas themselves or picking up and developing the dreams of other visionaries, inventors, and discoverers who came before them and were sometimes ahead of their time. Competitive conditions facilitate the implementation of breakthrough ideas, leading to the emergence of new technical systems.

When upgrading existing systems, if the changes being made are so significant that they go beyond the scope of previously conducted experimental and theoretical

DOI: 10.1201/9781003670575-9

substantiations, uncertainty may arise that negatively impacts safety. This impact becomes even more potentially dangerous when creating new types of systems. To minimize the resulting uncertainty and ensure safety in both cases, improvements to design methodology may be necessary. This aspect has been the leitmotif of the previous chapters of this book.

It remains for us to consider the general changes in the engineering design landscape that are taking place under the influence of the above-mentioned drivers of development. These changes, occurring over a significant period of time and across various industries, encompass:

- the object of design itself – the technical systems being created;
- the methodology used in their creation; and
- the regulation of design activities by the state in order to meet the need of society for safety.

The changes observed in the listed dimensions of the design landscape bring with them additional uncertainties. These are analyzed in the following sections.

9.2 Changes in the Technical Systems Being Created

In the process of technological development, two distinct trends can be identified:

1) an evolutionary approach, which consists of improving previous specimens of technical systems that serve as prototypes; and
2) the creation of new types of machines or facilities intended to perform new functions, using new working and control principles, as well as design solutions and materials that go beyond established traditions.

These trends apparently existed even before engineering design became a specialized field of human activity and a scientific and technical discipline (which did not occur until the late XIX century). Currently, after the rapid technological development in the XX century led to the emergence of not only fundamentally new types of engineered systems but also entire industries related to their life cycles, such as aviation, astronautics, nuclear energy, computing, and information technology, the first trend seems to dominate.

Nevertheless, the creation of new types of technical systems continues today, although this trend is not always realized in its purest form. Truly breakthrough ideas arise extremely rarely, and even seemingly pioneering achievements may have prototypes in the history of technology. In some cases, new technical systems are created by combining elements from different technological areas. One example is aerospace vehicles designed for suborbital transportation of people and cargo. They can combine the principles of ballistics, implemented in rocket technology, with the aerodynamic flight, which is the basis for the operation of air vehicles. Other examples include nuclear-powered spacecraft designed for long-distance missions and power plants placed in low Earth orbit to collect solar energy and transmit it to the Earth's surface.

Evolutionary design approach, i.e., the improvement of prototypes, is based on the use of known relationships between the functional characteristics of the system being developed (describing its purpose and performance) and its construction parameters (related to configuration, dimensions, material properties, and control). Due to such relationships and design solutions, substantiated not only theoretically but also confirmed by previous prototype operating experience, determining the construction parameters based on the specified functional characteristics is associated with less uncertainty than creating a new-generation technical system from scratch. However, uncertainty may also be present in the evolutionary approach, manifesting itself in the form of both aleatory and epistemic components.

Innovative design activities that go beyond the studied domains of the mentioned relationships or that use non-traditional combinations of heterogeneous elements in the created system generate significantly greater uncertainty. For example, when designing an aerospace vehicle, it is necessary to address issues such as the choice of:

- launch and landing scheme (vertical or horizontal with various acceleration options and the number of vehicle stages);
- the type and parameters of the propulsion system (rocket, turbojet, ramjet, combined);
- the range of flight velocities in the atmosphere (subsonic, supersonic, hypersonic);
- thermal insulation characteristics, etc.

When addressing these issues, it is necessary to take into account environmental aspects, including noise pollution, compatibility with existing ground infrastructure, airline routes, and so on.

Design considerations for a spacecraft with a nuclear propulsion system shall include the overall system configuration, taking into account biological shielding, the type of nuclear reaction, the method of generating the propulsion force, structural materials capable of withstanding extreme temperature loads over a long period of time, the final heat sink, and many others issues. Environmental aspects are less relevant for missions far from Earth, but consideration must be given to the loading and unloading of nuclear fuel, as well as the maintenance of the nuclear facility in extraterrestrial conditions.

A solar power plant in low-Earth orbit can be constructed as a thin-film or rigid multi-link structure, deployable from a folded, transportable state. It could function, for example, as a reflector, collecting sunlight and directing it toward Earth (Saprykin et al., 1998; Spirochkin and Saprykin, 2021). A space reflector can also be made as a liquid mirror, formed by the surface of a liquid that, when rotated, takes on a parabolic shape, ideal for emitting a beam of parallel rays. This method of creating mirrors is currently being studied for advanced space telescopes (Figliozzi, 2023; Durrani, 2025). Another option for constructing an orbital power plant involves using an array of photovoltaic cells and converters to transmit the harvested solar energy to Earth as a microwave beam (Blain, 2022).

In each of these construction options, the key issues are the formation of a large space structure and ensuring its required geometric shape, as well as controlling the orientation of mirrors or photovoltaic devices to receive maximum light flux.

The design and operation principles of such a power plant shall take into account the risks of collisions with space debris and meteorite impacts. Furthermore, the method of transmitting energy to the Earth's surface should be chosen based on an analysis of the impact of beam-transmitted energy on terrestrial ecosystems and flight safety, assuming the possibility of an aircraft entering the energy beam. All of the above issues have not yet been fully resolved and are being studied through experiments conducted both on Earth and in space during research missions. The uncertainty associated with them corresponds to the sources presented in Table 3.1.

Uncertainty is no less present in nuclear fusion projects, another rapidly developing area of science and technology (WNA, 2025; Midgley, 2025). The success of a single space or fusion experiment, confirming the choice of a promising implementation option for the technical system under design, can influence the further course of R&D and lead to local changes in the design landscape in the relevant area.

9.3 Changes in Design Methodology

9.3.1 Changes at the Level of General Methodology

The traditional design paradigm, presented in Sects. 1.2 and 2.1, is based on compliance with regulatory documents that explicitly or implicitly take into account statistical data obtained from experience in the creation and operation of prototypes, as well as on the use of scientifically substantiated risk management methods. At the same time, this paradigm does not exclude the reverse influence of post-design stages of the life cycle of a technical system on design decisions with the aim of improving them, but this influence is carried out without changes in the regulatory framework, which has significant inertia. The proper combination of these aspects ensures a level of safety that has been generally acceptable to society in previous decades of technological development. In Sect. 2.2, we discussed the shortcomings of this paradigm and the changes (shifts) needed to address them and meet modern, higher safety requirements necessitated by the high-profile disasters of recent years (some of which are mentioned in Sect. 1.2.2). The necessary shifts are achieved through the implementation of approaches such as in-service monitoring and diagnostics (Sect. 5.5) and life cycle management based on actual state (Sect. 6.2), or by applying the SPARS design concept (Sect. 6.4).

In our understanding, changing the design paradigm does not mean its complete replacement with some fundamentally different philosophy, but rather the careful incorporation of new concepts, theories, models, and methods into the set of previously used engineering tools. As a result, the entire field of engineering knowledge and the methodologies for its application (which is encompassed by the term "paradigm") or the "center of gravity" of this field shifts towards more adequate management of uncertainty and a higher level of safety. The main observed shifts of this kind, associated with the phases of evolution of the general design methodology and the qualities acquired by the designed systems, are presented in Table 9.1.

Table 9.1 Changes in the general design methodology.

Phase of evolution	*Characterization of the methodology*	*Acquired quality (qualities) of the system*
The traditional design paradigm	Function-oriented and statistically based approach to design	Acceptable (in previous decades) level of safety
Shift 1	Design for in-service monitoring and diagnostics (with a life cycle management perspective)	Improved functionality, efficiency and safety over a long service life
Shift 2	Design for safety in accidents	Increased safety in anticipated accidents
Shift 3	Design according to bionic principles	Smartness + proactivity + resilience = Increased safety in the face of unexpected dangers

The transition from one phase to another occurs over time, when a favorable, stable combination of necessary conditions (social needs, scientific and technological resources, and personal qualities of leaders) is formed. Each subsequent phase borrows valuable items from the previous one and shifts the boundaries of the existing design paradigm by introducing a new design methodology to achieve new sought-after quality (or qualities) in the systems being designed.

The methodology inherent to the traditional design paradigm existed and evolved throughout most of the XX century. In the case of innovative developments, it can be used without relying on statistics, which are often absent, focusing solely on the required system function and employing a trial and error approach. This version of the methodology can be quite expensive, but it is used by military systems developers, participants in the space race, and companies working in the field of AI.

The next phase in the evolution of design methodology, designated in Table 9.1 as "Shift 1", is perhaps the most characteristic for nuclear energy. NPPs are extremely expensive technical systems; their construction takes many years, and decommissioning and disposal of spent nuclear fuel, nuclear waste and irradiated structural materials pose significant challenges. Nuclear and radiation accidents are extremely rare, but mitigating the consequences of each one is costly, not to mention the negative public reaction that has led to the complete abandonment of nuclear energy in some countries. The combination of these factors determines the course towards designing NPPs with a long service life, as well as the need for in-service monitoring and diagnostics of nuclear facility elements. The target service life of modern NPPs is 60 years, with the possibility of further extension.

Life cycle management based on actual state (a methodology reflecting the realized perspective) includes, in addition to service life extension, relevant procedures for system maintenance, repair and replacement of its elements, as well as a possible change in the operating profile.

This methodology has also found application in aerospace engineering, for example in projects of long-lived and expensive space systems such as the Hubble and James Webb space telescopes, as well as the International Space Station. Considering a number of high-tech industries, the most intensive development of design methodology, ensuring the long service life of technical systems based on in-service monitoring and diagnostics, can be attributed to the turn of the XX and XXI centuries.

The beginning of the phase "Shift 2" can be traced back to the emergence of the crashworthiness concept in the US military aviation in the 1970s, aimed at ensuring the survivability of occupants of airplanes and helicopters in crashes. Some principles of crashworthiness were embodied in the design of the Soviet space shuttle *Buran* (see Sect. 6.4.2). Further application of this concept is related to ensuring safety in accidents in civil aviation, automobile and rail transport (Spirochkin, 2023). In the nuclear energy, similar principles are implemented through the use of structural elements such as:

- crushable supports, or pipe-whip restrains, intended to absorb kinetic energy released during ruptures of NPP pipelines and prevent their parts from whipping (KTA 3205.1, 2002; VERLIFE, 2013);
- a protective structure of the reactor building, or containment, capable of withstanding an airplane crash (IAEA, 2019, 2021);
- dampers, or impact limiters, installed on spent nuclear fuel transport containers to protect against extreme loads caused by impacts and drops (Johnson et al., 1999; IAEA, 2000).

However, design methodology aimed at ensuring safety in accidents is not limited to the concept of crashworthiness. Its scope is broader, and this methodology covers analysis of all possible accidents at the design stage. For nuclear facilities, these constitute the category of design basis accidents (see Sect. 3.1), postulated by regulatory documents, which must be taken into account. In this regard, another important direction of design for ensuring safety in accidents, which has emerged in recent time, is the creation of core catchers, which should work in the event of a severe accident accompanied by meltdown of the core and destruction of the reactor vessel (Svetlov et al., 2003).

In nuclear energy, another term that has an important meaning in the context of ensuring safety is also used: *beyond design basis accident*. It denotes "accident sequences that are possible but were not fully considered in the design process because they were judged to be too unlikely" (NRC, 2021). This term can be applied to any unexpected hazardous situation that a nuclear facility was not designed to withstand. Beyond design basis accident analysis, if carried out, is intended to "fully understand the capability of a design" (ibid) and determine the limits of changes in the key variables within which safety is ensured.

In other fields of engineering, this term is not used, although unforeseen hazardous situations are certainly possible. Consideration of this possibility, characterized by significant uncertainty, serves as an incentive to move to the next phase of development of the general design methodology, which we designated as "Shift 3". This shift is manifested, in particular, in the emergence of the SPARS design concept, presented in Sect. 6.4.

It should be noted that changes in design methodology are not directly related to changes in the technical systems being created, which were discussed in Sect. 9.2. These two dimensions of the design landscape are largely independent. We have already mentioned the circumstances that act as triggers and drivers for updating design approaches. Let us rephrase them: 1) the cost of errors made when applying old methodologies (including the unacceptable "cost" of accidents and catastrophes); 2) the accumulated scientific and technological potential for improving the productivity, efficiency, or safety of a technical system, which exceeds the capabilities of traditional methods and leads to their obsolescence; and 3) the presence of leaders capable of directing technological development.

9.3.2 Changes in Applied Methods

The changes described in Sect. 9.3.1 concerned the general methodology – a kind of philosophy – of design. However, even more significant changes are also occurring at the applied level. One example of such changes is the widespread transition, during the last decade of the XX century and the first few years of this century, to the creation of technical systems using CAD systems and the substantiation of their reliability using CAE systems based on finite element analysis. Another, more recent example is the replacement of experimental methods for studying the behavior of designed systems interacting with liquid or gas flows with computational fluid dynamics (CFD) methods.

At the same time, applied design methods are subject to the conservative influence of tradition, which manifests itself in the preferred use of standardized, practically proven structural elements and units in the creation of a system, a focus on manufacturing technologies that do not go beyond the framework of not only regulatory documents, but also established unwritten rules, and the control of possible errors in the computerized substantiation of design decisions by qualified specialists, as a rule, senior and most experienced individuals. The author recalls his early years in the aerospace industry as a young analysis engineer, when his superiors would check key computer calculation results against rough estimates derived from theoretical mechanics formulas using a slide rule. Adherence to best practices, which has become a tradition, certainly helps ensure reliability, but it also complicates the application of new structural materials, innovative technological processes, and advanced numerical methods capable of improving the performance of the system being developed. As a result, implementing such innovations in the development of safety-critical systems is sometimes possible only through partial modification of design practices and standards being in use, while maintaining the core traditions.

One example in this regard is the modification of a standard within the ASME Code governing the design of high-temperature nuclear reactors to include composite graphite and ceramic materials in their structures, which were previously made only of metals (ASME, 2019). The specifics of strength analysis of these new materials are considered in separate sections of the standard. Another example is the post-processing of the results from an elastic finite element stress analysis carried out for pressure vessels to obtain membrane and bending stresses consistent with traditional strength assessments using shell theory. This post-processing, called "linearization of stress results for stress classification", is provided for in the relevant sections of the ASME Code (ASME, 2007).

Clearly, advances in computer technology are significantly influencing changes in applied design methods. Computer simulations are playing an increasingly important role in substantiating design decisions, and, accordingly, the relative share of physical experiments is decreasing. However, this process can only be considered unambiguously positive when combined with the correct formulation of design problems, validation of numerical models, verification of software, and (still necessary) human control. If any of these conditions are not met, computer simulations can introduce uncertainty into the substantiation of design decisions (see source 6 in Table 3.1), which is difficult to assess.

9.4 Changes in Regulation

Above, in Sect. 9.3.2, we provided two examples of changes in design standards that reflect innovations in the construction of designed systems and in the applied engineering methods. In general, the role of standards is to preserve successful, proven design solutions with the aim of using them as recommendations when creating new systems. The use of standardized solutions increases the efficiency of the design process by minimizing the costs of researching problems that have already been solved once, ensuring the compatibility of the created system with existing technologies, and reducing the uncertainty associated with the operation stage. Thus, adherence to standards helps maintain the safety and performance of the designed system at the level achieved for prototypes. Standards are regulatory documents that are adopted by stakeholders, typically on a voluntary basis, due to the recognized value of these documents.

It should be noted that standards belong to the lower level in the hierarchy of the national regulatory framework governing the safety of the use of technical systems. The upper level is formed by legal documents establishing the requirements that technical systems must meet to obtain a state license for their creation and operation. In the United States, such documents constitute the Code of Federal Regulations (CFR), and in Russia, they are Federal Laws governing specific industries, such as the 'Air Code of the Russian Federation' (No. 60-FZ, 1997), 'On the use of atomic energy' (No. 170-FZ, 1995), and other regulatory legal documents adopted by legislative authorities. In nuclear energy, the latter category is represented by the 'Federal regulations and rules in the field of atomic energy use' – a series of documents with the index 'NP', starting from NP-001-15.

Since upper-level regulatory documents establish safety requirements that must be met regardless of the specific implementation of a technical system, they typically contain safety criteria expressed in general terms and do not specify any methods, design features, or variable values that can ensure that these criteria are satisfied. However, this was not always the same situation – the original design rules developed in the early stages of various industries contained technical details, and clearing legal regulatory documents from these details can be a lengthy and complex process. In some cases, it is still not completed. An example of excessive detailing that carries the imprint of the past and potentially complicates progressive design decisions is the use of acceleration values in airworthiness standards for transport-category airplanes to regulate the required strength of structural elements in the design case "emergency landing" (Sect. 7.2.2). The lack of invariant safety criteria that can account for a wide range of aircraft configurations, structural materials, and landing conditions creates uncertainty that negatively impacts safety.

Nevertheless, regulatory changes that facilitate innovation are underway. These changes include, for example, the possibility of applying the equivalence principle, according to which the safety of an aircraft with an unusual design and (or) made from new materials can be substantiated for the event of emergency landing by demonstrating that the safety level is equivalent to that of existing prototypes with a conventional configuration manufactured using well-studied materials (Federal Register, 2015). A generally favorable trend in aerospace regulation is that licensing authorities are moving away from setting design and manufacturing rules and focusing solely on the level of safety that newly developed aerospace systems must meet. This trend manifested itself, in particular, by the introduction of the 'Streamlined Launch and Reentry Licensing Requirements' by FAA in 2020 (Federal Register, 2021).

In nuclear energy, changes in regulation occur primarily through the development of new regulatory documents for each new type of nuclear facilities, taking into account the specifics of their design and the physics of working processes. The construction of fast neutron reactors with sodium or lead coolant in Russia generally complies with the safety principles set out in the national top-level regulatory documents (PNAE G-7-002-86; NP-001-15), therefore, changes to individual elements and operating modes in a newly designed facility of this type are accompanied mainly by updating the standards, i.e., adjusting the lower level of regulation. However, the design of molten salt reactors goes beyond these principles; in particular, they have fewer physical barriers to prevent the release of ionizing radiation into the environment. Therefore, the creation of such reactors may require updates to regulations at both levels. These updates will likely affect a number of regulatory provisions, including the accepted factors of safety.

9.5 Concluding Remarks: New Challenges of Uncertainty

Let us summarize the consideration of many topics and issues in this book. A wide range of approaches to minimizing uncertainty in the design of technical systems was presented and discussed – both well-known methods used for many years and new concepts currently being developed and tested. More sophisticated of these methods expand the possibilities for reducing uncertainty, including its epistemic component, by transferring some of the uncertainty management activities from the design stage to operation. As a result, safety of the created technical system can be ensured at a level higher than that achievable within the framework of the traditional design paradigm with its inherent a priori, statistically based approach to uncertainty.

However, it would be presumptuous to expect complete control over such a complex and multifaceted phenomenon as uncertainty, which is subjective in nature and, moreover, evolves along with changes in the world around us, as well as with our reaction to these changes. This evolution manifests itself in new challenges of uncertainty, among which the following should be noted:

- changes in the natural environment, in particular climate change;
- the emergence of new threats due to the expansion of the human-created technosphere and its internal transformation;
- the introduction of AI, the consequences of which are now difficult to imagine.

Climate change is altering the characteristics of meteorological impacts on technical systems. Changes in atmospheric temperature and stratification, wind directions and speeds, the frequency and intensity of hurricanes, the intensity of ocean waves (including tsunamis), and other natural phenomena can directly affect safety of aircraft and ships, energy and industrial facilities, and other infrastructure. These changes are not yet precisely determined.

New threats associated with the expansion of the technosphere include, for example, the increasing amount of orbital debris, which increases the risk of its collisions with spacecraft, as well as the danger of fragments of this debris falling uncontrolled to Earth. Such a fall threatens to destroy ground objects and aircraft in flight, not to mention the possibility of loss of life.

The internal transformation of the technosphere involves replacing old technologies with new ones that their creators consider more efficient. This transformation may carry the risk of insufficient substantiation for the safety of new technologies, especially those developed under conditions of financial pressure and market competition. Unfortunately, cost savings sometimes come at the expense of safety. Technological advances can also lead to incompatibility between new technologies and existing, long-standing systems, as well as a loss of knowledge about older technologies. The impact of these aspects on the future of humanity has not yet been sufficiently studied.

The consequences of the introduction of AI is particularly difficult to predict when it occurs in the control of safety-critical technical systems and evolves against a backdrop of declining use of human natural cognitive abilities.

These are just a few of the challenges of uncertainty that we are currently observing and trying to understand. Along with them, there are unforeseen threats that lie beyond our comprehension. We should not count on a guaranteed safe existence. Ensuring safety in the creation of future generations of technical systems requires not only dedicated engineering efforts but also attention to ongoing changes in the world, even those that initially seem insignificant, and a readiness to address emerging challenges. The author hopes that the methods presented in this book will remain applicable to solving new problems, and that some of the ideas expressed will contribute to the further development of the philosophy of design.

References

ASME (2007) An International Code. 2007 ASME Boiler & Pressure Vessel Code. Section VIII. Division 2. Alternative Rules. Rules for construction of pressure vessels. American Society of Mechanical Engineers (ASME), New York.

ASME (2019) An International Code. 2019 ASME Boiler & Pressure Vessel Code. Section III. Rules for construction of nuclear facility components. Division 5 – High temperature reactors. American Society of Mechanical Engineers (ASME), New York.

Blain, L. (2022) Caltech's space solar project prepares for its first orbital prototype. *New Atlas*, October 18, 2022. https://newatlas.com/energy/caltech-space-solar-power-project. Accessed November 23, 2025.

Durrani, I. (2025) Explaining liquid mirror space: a reflective revolution in observational cosmology. ResearchGate, August 05, 2025. DOI: 10.13140/RG.2.2.21196.73601. https://www.researchgate.net/publication/394310065_Explaining_Liquid_Mirror_Space_A_Reflective_Revolution_in_Observational_Cosmology. Accessed November 24, 2025.

Federal Register (2015) Vol. 80, No. 107/Thursday, June 4, 2015/Notices, pp. 31946–31948. Department of Transportation. Federal Aviation Administration. Aviation Rulemaking Advisory Committee; Transport airplane and engine issues; New task. https://www.faa.gov/regulations_policies/rulemaking/committees/documents/media/TAEcdeT1-06042015.pdf. Accessed November 30, 2025.

Federal Register (2021) Streamlined launch and reentry license requirements. A Rule by the Federal Aviation Administration on 03/09/2021. Federal Register. *The Daily Journal of the United States Government*. https://www.federalregister.gov/documents/2021/03/09/2021-04068/streamlined-launch-and-reentry-license-requirements. Accessed November 30, 2025.

Figliozzi, G. (2023) What is the fluidic telescope? NASA, April 27, 2023. https://www.nasa.gov/science-research/astrophysics/what-is-the-fluidic-telescope. Accessed November 24, 2025.

IAEA (2000) Multi-purpose container technologies for spent fuel management. IAEA-TECDOC-1192. International Atomic Energy Agency, Vienna.

IAEA (2019) IAEA Safety Standards Series No. SSG-53. Design of the reactor containment and associated systems for nuclear power plants. Specific safety guide. International Atomic Energy Agency, Vienna.

IAEA (2021) IAEA Safety Standards Series No. SSG-68. Design of nuclear installations against external events excluding earthquakes. Specific safety guide. International Atomic Energy Agency, Vienna.

Johnson, R.A., McInnes, I.D., Quinn, R.D. and Temus, C.J. (1999) Impact limiter for spent nuclear fuel transportation cask. European Patent Specification. EP 0 674 801 B1. Date of publication: 08.09.1999.

KTA 3205.1 (2002) Sicherheitstechnische Regel des KTA. Komponentenstützkonstruktionen mit nichtintegralen Anschlüssen. Teil 1: Komponentenstützkonstruktionen mit nichtintegralen Anschlüssen für Primärkreiskomponenten in Leichtwasserreaktoren. Anhang D. Der Kerntechnische Ausschuss (KTA). Salzgitter, Deutschland.

Midgley, E. (2025) Fusion energy in 2025: six global trends to watch. IAEA / News, October 28, 2025. https://www.iaea.org/newscenter/news/fusion-energy-in-2025-six-global-trends-to-watch. Accessed February 01, 2026.

No. 60-FZ (1997) 'Vozdushnyi Kodeks Rossiiskoi Federatsii' ot 19.03.1997 No. 60-FZ (Air Code of the Russian Federation), with amendments and additions, entered into force on November 26, 2025. https://www.consultant.ru/document/cons_doc_LAW_13744. Accessed November 30, 2025 (in Russian).

No. 170-FZ (1995) Federalnyi Zakon 'Ob ispolzovanii atomnoi energii' ot 21.11.1995 No. 170-FZ (The Federal Law 'On the use of atomic energy', edition dated December 26, 2024. https://www.consultant.ru/document/cons_doc_LAW_8450. Accessed December 03, 2025 (in Russian).

NP-001-15 (2015) Federalnye normy i pravila v oblasti ispolzovaniya atomnoi energii. Obshchie polozheniya obespecheniya bezopasnosti atomnykh stantsii (Federal regulations and rules in the field of atomic energy use. General provisions for ensuring safety of nuclear power plants). Federal Environmental, Technological and Nuclear Supervision Service of Russia, Moscow (in Russian).

NRC (2021) Beyond design basis accident. U.S. NRC – United States Nuclear Regulatory Commission, March 09, 2021. https://www.nrc.gov/reading-rm/basic-ref/glossary/beyond-design-basis-accidents. Accessed November 28, 2025.

PNAE G-7-002-86 (1989) Normy raschyota na prochnost oborudovaniya i truboprovodov atomnykh energeticheskikh ustanovok (*Rules for strength analysis of equipment and pipelines of nuclear energy facilities*). Gosatomenergonadzor of the USSR. Energoatomizdat, Moscow (in Russian).

Saprykin, O.A., Spirochkin, Y.K., Kinelev, V.G. and Sulimov, V.D. (1998) Computer aided optimal design of space reflectors and radiation concentrators. *AIP Conference Proceedings*. Volume 433, Issue 1, pp. 460–468.

Spirochkin, Y.K. and Saprykin, O.A. (2021) Kosmicheskii polyot pod solnechnym parusom (Space flight under a solar sail). *AviaSafety.ru*, January 6, 2021 (in Russian). https://aviasafety.ru/36409. Accessed November 23, 2025.

Spirochkin, Y. (2023) *Human factors and design*. Springer, Singapore.

Svetlov, S.V., Bezlepkin, V.V., Kukhtevich, I.V., et al. (2003) Core catcher for Tianwan NPP with WWER-1000 reactor. Concept design and justification. *Proceedings of the 11th international conference on nuclear engineering* (ICONE-11), Tokyo (Japan), 20-23 April 2003.

VERLIFE (2013) Guidelines for integrity and lifetime assessment of components and piping in WWER nuclear power plants. IAEA, Vienna. Appendix F – Component and piping supports (Contributors: Obushev A., Spirochkin Y.).

WNA (2025) Nuclear fusion power. World Nuclear Association / Information Library / Current and Future Generation, June 5, 2025. https://world-nuclear.org/information-library/current-and-future-generation/nuclear-fusion-power. Accessed February 01, 2026.

Index

About the Author

Yuri Spirochkin is an engineer, researcher and author with extensive experience in the aerospace and nuclear power industries. He graduated from the N.E. Bauman Moscow Higher Technical School (now the N.E. Bauman Moscow State Technical University). At the beginning of his career, he took part in the design of the Soviet space transport system *Energia-Buran*. He created the mathematical model of the *Buran* shuttle and carried out research of its possible emergency landing. In the early 2000s, he was involved in R&D in the field of bioengineering. He later worked in the Russian nuclear power industry, participating in rulemaking, design and life cycle management of nuclear power plant equipment and piping. For several years he was a visiting expert to the International Atomic Energy Agency. Currently he serves as an expert at the International Consultancy and Analysis Agency 'Aviation Safety', focusing on aerospace engineering issues. His publications include a chapter in an edited volume, three books and numerous articles, as well as several patents.